Introduction *to* Geodesy

Introduction *to* Geodesy

Clair E. Ewing *and*
Michael M. Mitchell

ELSEVIER

NEW YORK   OXFORD   AMSTERDAM

536
E 95

ELSEVIER NORTH-HOLLAND, INC.
52 Vanderbilt Avenue, New York, N. Y. 10017

ELSEVIER SCIENTIFIC PUBLISHING COMPANY
335 Jan Van Galenstraat, P.O. Box 211
Amsterdam, The Netherlands

Standard Book Number 444-00055-0

Library of Congress Card Number 69-11221

Third Printing, 1976

Printed in the United States of America

Contents

Preface

This book was written to fill a specific need and not simply as an academic exercise. Recently, the authors taught introductory courses in geodesy to students who had either a bachelor of science or a master of science degree in some technical field, but who were being introduced for the first time to the field of geodesy. In preparing for the course, it became clearly evident that a modern introductory textbook and reference book is vitally needed. At one end of the spectrum of available literature is the classical work of Dr. W. A. Heiskanen and Dr. F. A. Vening Meinesz, "The Earth and Its Gravity Field," an outstanding book, but too far advanced for students exposed to geodesy for the first time. At the opposite end is "Geodesy for the Layman," a highly informative and unique document, written for the United States Air Force Aeronautical Chart and Information Center by Richard K. Burkard to explain some of the basic principles of the science. Between these two extremes there is a gap; nothing is available to answer this need. As a result of this dilemma, it became necessary to write and reproduce for the students a set of lessons designed for the specific task at hand.

Responding to encouragement to publish these lessons in textbook form, the authors expanded the original material to include recent knowledge, particularly in the area of satellite geodesy. The result is this book, prepared by the authors as private individuals and not as representatives of the United States Air Force. As the title suggests, the goal of the book is to provide the reader with a means for understanding the basic concepts of the science. It is intended for the student at the upper undergraduate or lower graduate level when used as a textbook, or for individuals with a comparable education when used as a reference book. It assumes that the reader has a familiarity with mathematics including calculus. Mathematical derivations are employed only when necessary for the proper understanding of a process; where their development is of questionable value, or more properly belongs in an advanced textbook, only the results are presented and explained. If the authors have achieved their objective, this book should yield a depth of information sufficient both to provide the reader with a fundamental knowledge of geodesy, and, if he so desires, to equip him for advanced study. To assist the reader who wishes to pursue the subject, current sources of additional information are given.

The book is arranged in what the authors believe to be a logical approach to learning. First comes classical geodesy dealing with the properties of the ellipsoid, calculation of triangulation, and computations on the ellipsoid. The next chapter is devoted to geodetic astronomy, or the determination of latitude, longitude, and azimuth. Then coordinate systems and their relationships are discussed. Electronic surveying, which has come into its own within the past twenty years, is explained. Finally physical geodesy, leveling, satellite geodesy, and adjustment computations are explained.

On the premise that the difference between the technician and the professional is an understanding of the theory underlying the processes, examples are used only when necessary to illustrate a difficult point. This book is not a manual on doing geodesy because suitable field and computational manuals are available from the U.S. Coast and Geodetic Survey and other organizations engaged in geodetic activities. It is a book written in a simple, informal manner, designed to introduce the reader to the fascinating science which concerns the size and shape of the earth, its gravity field, and man's computations on its surface and the surrounding space.

CLAIR E. EWING
MICHAEL M. MITCHELL

Lompoc, California, and *Washington, D.C.*
March 1969

Chapter 1   Introduction

Geodesy is that branch of applied mathematics which determines, by observation and measurement, the size and shape of the earth, the coordinates of points, the lengths and directions of lines on the earth's surface, and the variations of terrestrial gravity. Geodesy is generally considered to have two separate parts. The first, geometrical geodesy, concerns the size and shape of the earth, the intercontinental ties among the land masses of the earth, and the determination of positions, lengths of lines, and azimuths; as the name implies, it has to do with the geometry of the earth. The second, physical geodesy, concerns the gravity field of the earth, or the direction and magnitude of the physical force which links the earth to objects on its surface and in surrounding space. A study of gravity is useful in determining the shape, but not the size, of the earth. Although these two parts of geodesy may be studied separately, in practice they are closely interwoven.

No specific date in history marks the beginning of geodesy. As soon as primitive man reached the stage in his development when he began to reason and wonder, certainly he must have questioned the extent of his domain. When first he saw the shadow of the earth projected on the moon during a lunar eclipse, he must have reasoned that the earth was a circular disk. As he became a sea-going creature and observed the ships sail over the horizon, he obviously deduced that the earth had curvature in all directions. Piecing these two bits of information together, he could have concluded very early in his existence that the earth was spherical in shape. At that time, he had no tools with which to prove his hypothesis or determine the size of his domain.

The problems of geometrical geodesy have plagued mankind throughout the centuries. The size and shape of the earth were interesting research problems to the scientists of several thousand years ago. It is extremely interesting to note the first scientific method used by the Greek scholar Eratosthenes about 220 B.C. Eratosthenes was Keeper-of-the-Scrolls for Egypt, with headquarters in Alexandria. His job might be compared with that of the Chief Librarian of Congress in this country. He often pondered the problem of determining the true size and shape of the earth. Finally, a solution came to his mind. Figure 1–1 shows the geometry of the problem. There were three elements that he needed. The first was a precisely measured baseline in a north-south direction. This he established between Alexandria on the north and Syene (or Aswan, as it is known today) on the south. In order to accomplish this measurement, Eratosthenes used a camel. Being a man of science, he doubtless refused to use an ordinary run-of-the-stable camel, but insisted on a well-calibrated camel. After repeated calibration runs over the sands of the Nile, he knew down to the last camel's length just how far his camel would go in the time it took the sand to empty from one side of the hourglass to the other. After making his measurement between Alexandria and Syene, he pronounced the baseline to be 5000 stadia in length.*

The second element of the problem was to have the southern end of the baseline lie exactly on the Tropic of Cancer. This was necessary so that the sun's rays on the longest day of the year (the summer solstice) would be directly overhead and thus intersect the center of the earth if they could be prolonged. Syene's position on the Tropic of Cancer was confirmed by Eratosthenes when he looked down a deep well on the longest day of the year and saw that the sun's rays reached the bottom; he knew then that the sun was directly overhead.

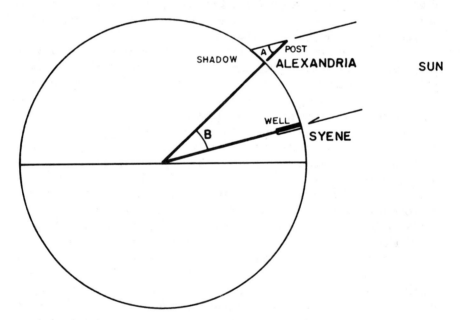

Figure 1–1.   Method of Eratosthenes.

---

* A stadium is about $\frac{1}{10}$ of a modern nautical mile.

The third and last element was a post in Alexandria. As the next summer approached, the length of the noon shadow cast by the post gradually shortened until finally, on the day of the summer solstice, it reached minimum length. The angle A was simply the angle whose tangent is the length of the shadow divided by the height of the post, and proved to be 7.2°. From simple geometry, with the sun's rays quite properly considered to be parallel, angle A and angle B are equal because they are alternate interior angles. Also, by simple proportion,

$$\frac{B}{360°} = \frac{5000 \text{ stadia}}{C_E}$$

where C_E is the circumference of the earth. Since C_E is the only unknown in the equation, the problem was easily solved. The circumference came out to be 250,000 stadia, about 16% too large in the light of modern knowledge of the length of the stadium and the size of the earth [1].

So in 220 B.C., over 1700 years before Columbus *thought* the earth was a sphere, and nearly 1800 years before Magellan *proved* the earth was a sphere by sailing around it, a Greek scholar not only thought the earth was spherical and proved it, but also determined its size to a high degree of accuracy.

The next such arc measurement was done by Poseidonius (135–50 B.C.), who used a baseline between Rhodus and Alexandria. This length had been determined on the basis of a ship's sailing time between the two cities. The earth's radius as determined by Poseidonius was 11% too large [1].

The caliph Abdullah al Mamum (A.D. 786–833) supervised the next recorded measurement of the earth's dimensions. This work was done on the Zinjar plateau close to Baghdad. Wooden rods were used to measure the baseline and the angles were made with greater precision than in the earlier determinations. The final solution showed a value 3.6% too great [1].

These early determinations are of interest chiefly because of the method used. Except for more precise instrumentation, the recent determinations have used essentially the same techniques. Precision theodolites are now used to measure the astronomical data, while chains of triangles, expanded from accurately measured baselines, have replaced the camel, the ships, and the wooden rod to provide the length of the arc.

Had there been worldwide communication networks at the time of Eratosthenes, his work might have become universally known and accepted. As it was, an assortment of ideas of the size and shape of the earth persisted through the centuries. One interesting concept was that of Theophrastus Bombast von Hohenheim, better known as Paracelsus,

the German medic. He believed the earth to be a hollow sphere, popu-
lated inside with gnomes. The term "gnomonic projection" lingers today from that concept; it is a projection on a tangent plane as observed
by a gnome at the center of the earth.

In recent times, there have been dozens of determinations of the
dimensions of the earth, or, more properly, of the reference ellipsoid,
which is a mathematical surface most nearly approximating the shape
of the earth. The most significant development in arc measuring is the
classical work of Hough and others of the U.S. Army Map Service
(AMS). This determination has become known as the Hough ellipsoid,
or sometimes as the 1956 AMS ellipsoid. The significant fact in this
recent determination is the length and geographical dispersion of the
arcs used. Prior to the Hough determination, none except the Krasovskii
solution had used individual connected arcs longer than 50° of central
angle. Hough used four arcs, two of which are the longest of their kind
ever used—about 100° in length. These four arcs are shown in Fig. 1–2:
a meridional arc extending from South Africa to Scandinavia, a meridional arc extending from Chile to Canada, a parallel traversing the
United States, and a parallel extending from Western Europe to
Siberia.

Another consideration of geometrical geodesy is the accuracy of the

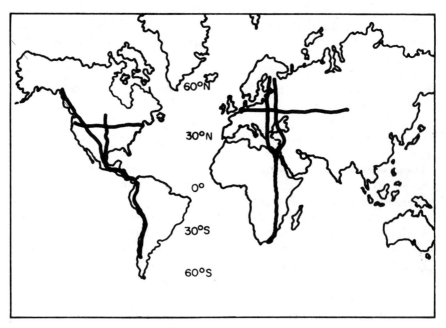

Figure 1–2. Arcs used in Hough's determination of the figure of the earth.

intercontinental survey ties. The distances between the several land masses of the earth are subject to errors which vary from a few hundred to several thousand feet. These intercontinental ties are made in a variety of ways, including optical, electronic, and celestial methods. The optical method is limited to line of sight between the two land masses, or observation upon a common object, such as a flare, from points on both land masses. This can be used only for such tasks as performing the tie between England and the Continent. Electronic devices such as Shoran (short-range navigation) and Hiran (high-precision Shoran) have accomplished ties like the one across the Mediterranean Sea between Africa and Europe and, by leap-frogging from Norway to Scotland to the Faroes to Iceland to Greenland to Canada, have completed the tie between Europe and North America.

Other broad expanses of water have not had such convenient stepping stones. For long ties, a number of astronomical means have been used, such as the solar eclipse, star occultations, the moon camera, and the satellite. All these methods employ geometrical relationships existing between observation stations on the land masses and a celestial body, which in the case of the satellite method is a man-made one.

By these different methods, intercontinental ties can be made. The relative position of all continents will be known much more accurately within the next few years.

Still another facet of geometrical geodesy is the determination of azimuths, lengths of lines, and geodetic coordinates of specific points. These are the final products of work-a-day geodesy so essential to mapping and charting, missile and space operations, and other geodetic activities.

In the realm of physical geodesy, the problem is to determine the gravity field of the earth and apply the information to useful geodetic purposes. Gravity on the earth is made up of three elements. Two of these vary in a regular mathematical manner over the earth, while the other changes in an irregular manner from point to point. The first of these is the attraction between the mass of the earth and a mass on the surface of a static earth. Just as Joshua said, "Sun, stand thou still," let us assume for a moment that the earth's rotation could be stopped. Gravity then would merely follow Newton's universal law of gravitation, varying directly as the product of the masses and inversely as the square of the distance from the center of the earth. It follows, since the earth is flattened at the poles, that gravity is greater at the pole than at the equator.

Next, allowing the earth again to spin on its axis and examining the effect of centrifugal acceleration, we can see that this factor tends to counteract gravity since its effect is outward. It may be said that centrifugal acceleration is antigravity. It is obvious that the greatest effect is at the equator, and that there is absolutely no effect at the pole since the pole is on the axis of rotation. Between the equator and the pole, centrifugal acceleration varies in a regular mathematical manner. Lumping the static effect with the rotational effect produces an equation which says that gravity anywhere on the earth's surface varies only as a function of ϕ, the latitude. All that is necessary to solve the problem is to plug the latitude into the equation. Considering only the regular variation of gravity, there is a difference of 1 part in 200 between pole and equator. A 200-pound man at the pole would weigh 199 pounds if transported to the equator.

However, there is a disturbing factor which complicates the problem. The earth is not homogeneous. Its crust is thick in places and thin in others. The thickness varies between about 6 and 60 kilometers, with an average near 35 kilometers. One explanation for the variation in crustal thickness is the roots-of-the-mountains theory. This theory postulates that the crust floats in a sea of dense magma which is probably in a plastic state. Analogous to this is a raft of logs floating down the river. From our own experience, if we see a group of logs of equal density in the water and observe one with a large mass above the water line, we know at once that it must have a large mass below the water line to provide sufficient buoyancy to sustain the large mass which we see. Also, if we see a tiny stick above the water, we decide that only a small amount is below the water line. So it is with the crust. A large mountain mass will sink deeply into the magma until it finds enough buoyancy to support it. This depression into the magma is called a mountain root. A smaller mountain will have a smaller root. An ocean basin or ocean trench will produce what is known as an antiroot.

Not only are there excesses and deficiencies in the volume of earth material but there is a variation in its specific gravity as well. The specific gravity may range from 2.67 for granite to 3.27 for magma, and even to 1.03 for sea water. Not only is there a change in gravity *intensity* but there is also a change in gravity *direction*. If a mass were suspended in the vicinity of a mountain, it would be deflected away from the vertical because of the proximity of the massive mountain on the one side and the absence of material on the other. The maximum amount of this deflection on the earth's surface is about 85 seconds of arc.

These variations from normal gravity are called anomalies. They distort the shape of the earth from a regular mathematical surface to an irregular shape whose definition requires an infinite number of parameters. This irregular shape is classically defined as mean sea level. One task of the geodesist is to find the best-fitting geometric figure which most closely approximates the true size and shape of the earth. Another result of gravity anomalies is the deflection of the direction of gravity away from geodetic vertical. Astronomical positions must be corrected for this deflection in order to attain true geodetic positions.

The change in gravity between points on the earth's surface is a problem in many fields of endeavor. A few years ago, Dr. W. A. Heiskanen analyzed its effect upon Olympic records. The results are startling. Let us say that in 1952 a shotputter heaved the 16-pound ball an even 60 feet in Helsinki. Let us further assume that in the next four years he maintained peak form so that at Melbourne in 1956 he was able to impart exactly the same energy to the shot as he did four years earlier in Helsinki. When the judges measured, they would have found he had achieved a distance of 60 feet 1½ inches because of the difference in gravity. The contrast is even more dramatic in the case of the javelin throw: the difference would be 6 inches. A broad jumper would gain ¾ inch. All these objects—the shot, the javelin, and the broad jumper—are bodies traveling through space, subject to universal physical laws; in short they are ballistic.

The combined problems of geometrical geodesy and physical geodesy have been brought sharply into focus since the advent of the missile and space age. For instance, a missile may be hundreds of miles above the earth, showing only as a bright pinpoint on a telescopic photograph and traveling at suborbital velocity, but it is still linked by physical laws to the earth below. If a missileer is to hit his target, he must know the earth's size and shape, the location of the launch pad and targets, and the gravity field of the earth.

These are the green years of geodesy. With the advent of airborne and satellite instrumentation, the science has been lifted from its earth-bound environment. Along with advancements in geodesy, rapid progress is also being made in all other earth sciences. The American Geophysical Union lists ten scientific fields under the general heading of geophysics. In addition to geodesy, they are planetary sciences, aeronomy, meteorology, hydrology, oceanography, seismology, volcanology, magnetism, and tectonophysics. All these scientific fields are to some degree related to geodesy. From the ever-increasing quantities of new and significant data valuable information about the earth will be learned.

REFERENCE

1. W. A. Heiskanen, Classroom Lecture Notes, The Ohio State University, Columbus, Ohio, September, 1953.

Chapter 2 The Ellipsoid of Revolution

The end product of geodesy is a set of coordinates, the length and azimuth of a line, or other meaningful data. While it is necessary to make observations and measurements on or near the physical surface of the earth, it would be quite impossible to perform detailed and extensive computations on a surface whose definition requires an infinite number of parameters.

A possible surface upon which to compute is mean sea level, or its extension into the land areas of the earth if channels were cut through the continents so that the sea could enter and come to an undisturbed and free state of equilibrium. We must assume these channels to be very narrow so that the quantity of water removed from the oceans would be negligible. This undisturbed water surface is known as the bounding equipotential surface of the earth or, more commonly, the geoid. It is everywhere normal to the direction of gravity. Because of mass excesses and deficiencies within the earth, the shape of the geoid is irregular and can be determined only approximately. This fact was mentioned briefly in the introduction and will be discussed in detail in Chapter 8. Like the physical surface, the geoid is unsuitable as a mathematical model for computations because it is defined by an infinite number of parameters.

If the earth were only slightly flattened, a sphere would be an ideal mathematical surface. It is a simple figure defined by only one parameter, its radius. As it is, the earth's departure from a true spherical shape is too great and so the thought of a sphere must be discarded.

Some other mathematical figure must be adopted. It must be simple enough so that computations are not overly difficult, but must nowhere depart from the true figure of the earth by an amount which will give intolerable errors in the results. This figure is the ellipsoid of revolution, generally referred to simply as the ellipsoid; it is produced by rotating an ellipse about its minor axis, with the major axis generating the equatorial plane. This ellipsoid of revolution approximates an oblate spheroid and the terms "ellipsoid of revolution" and "spheroid" are used interchangeably.

Much has been said in the past hundred years about the triaxial ellipsoid. This figure has three axes of unequal dimensions. In addition to the polar axis being shorter than the equatorial radius, as in the case of the simple ellipsoid of revolution, the equatorial plane itself is an ellipse with two axes of unequal length. This is an inconvenient figure to use but it is of academic interest because of its persistent mention for over a century. It will be given further treatment later in this chapter.

In the past few decades, data derived from gravity observations and also from satellite tracking have shown that the southern hemisphere is slightly larger than the northern, lending credence to the theory of the "pear-shaped earth." If a radius vector were drawn from the geocenter to a point at 45° north latitude, and another to a point at 45° south latitude, the latter would be longer by about 10 to 15 meters. This pear shape is known as an ovaloid. Putting all this information together indicates that the earth is an oblate triaxial ovaloid. The authors defy even the most gifted mathematician to produce a simple equation or procedure for computing upon this complicated mathematical surface.

We must accept the fact that the geoid is undulating, or warped, and try to find the ellipsoid of revolution which most nearly approximates the geoid with the least departure therefrom. It is analogous to the work of a civil engineer who builds a road through hilly country and seeks to find the route where the cuts and fills most nearly balance. The ellipsoid of revolution is a sensible mathematical model upon which to compute and is the cornerstone of geometrical geodesy.

Figure 2–1 shows the relationship among the physical surface, the geoid, and the ellipsoid of revolution at a small section of the earth's

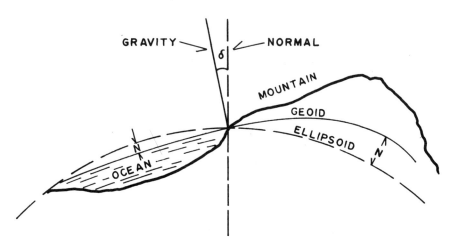

Figure 2–1. Relationship among the physical surface, the geoid, and the ellipsoid.

surface. At sea, the geoid coincides with the free, unbounded water surface. Under the continents it rises above the ellipsoid. The vertical separation of the geoid from the ellipsoid is known as the geoid height and designated as N. The geoid height may be positive or negative, depending on whether it is above or below the ellipsoid. The angle between the ellipsoid and the geoid at any point is called the deflection of the vertical and designated as δ. It is also the angle between the direction of gravity and the normal to the ellipsoid at any specific point, which is the same thing.

Because the ellipsoid of revolution has been adopted as the closest reasonable approximation to the figure of the earth, it is necessary to have a fundamental understanding of its mathematical properties. The ellipsoid is generated by rotating an ellipse about its minor axis. An ellipse can be defined in several ways. Considered as a plane curve, an ellipse is the locus of a point which moves so that the sum of its distances from two fixed points, called foci, is constant and greater than the distance between the two points.

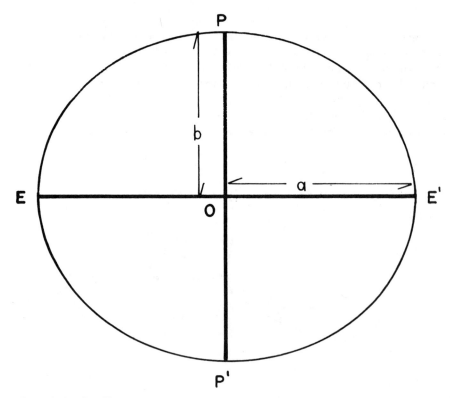

Figure 2–2. The ellipse.

Figure 2–2 is an ellipse. PP' is the minor axis, has length $2b$, and is the axis of rotation for generating the ellipsoid. EE' is the major axis, has length $2a$, and generates the equatorial plane when the ellipse is rotated about its minor axis.

A basic principle of the ellipse is that a circle of radius equal in length to the semimajor axis and with P (or P') as the center cuts the major axis EE' at F and F', points of special interest called foci. Figure 2–3 shows this construction. The distance OF then becomes $(a^2 - b^2)^{1/2}$.

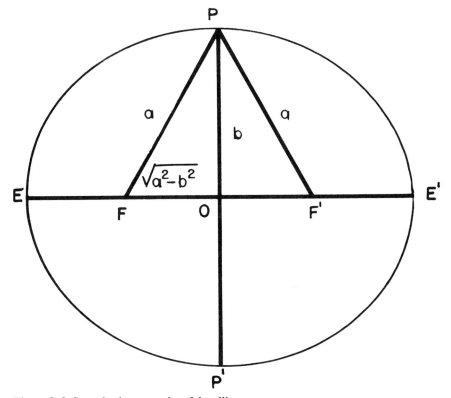

Figure 2–3. Some basic properties of the ellipse.

The flattening of the ellipse is designated f and defined as

$$f = \frac{a - b}{a} = 1 - \frac{b}{a}. \tag{2-1}$$

Occasionally, the reciprocal of the flattening is used in order to avoid fractions.

The first eccentricity of the ellipse is known as e. It is the distance between the center of the ellipse and a focus divided by the semimajor axis,

$$e = \frac{OF}{OE} = \frac{(a^2 - b^2)^{1/2}}{a}.$$

For practical purposes in computations, and in order to avoid the radical, the square of the eccentricity is used,

$$e^2 = \frac{a^2 - b^2}{a^2} = 1 - \frac{b^2}{a^2}. \tag{2-2}$$

The second eccentricity of the ellipse is known as e'. It is the distance between the center of the ellipse and a focus divided by the semiminor axis,

$$e' = \frac{OF}{OP} = \frac{(a^2 - b^2)^{1/2}}{b},$$

or

$$e'^2 = \frac{a^2 - b^2}{b^2} = \frac{a^2}{b^2} - 1.$$

The second eccentricity, e', is seldom used in geodetic computations and will not be discussed further in this book. When eccentricity is mentioned, it will be the first eccentricity, e.

The eccentricity and flattening have a definite mathematical relationship. From Eq. (2-2) we get

$$\frac{b^2}{a^2} = 1 - e^2, \quad \text{or} \quad \frac{b}{a} = (1 - e^2)^{1/2}. \tag{2-3}$$

Substituting the value of b/a from Eq. (2-3) into Eq. (2-1) produces

$$f = 1 - (1 - e^2)^{1/2},$$

or

$$(1 - e^2)^{1/2} = 1 - f.$$

Squaring this equation gives

$$1 - e^2 = 1 - 2f + f^2,$$

from which

$$e^2 = 2f - f^2. \tag{2-4}$$

Because f^2 is quite small, we can say approximately that

$$e^2 = 2f. \tag{2-5}$$

The parameters a, b, f, e, and e' are the principal parameters of the ellipsoid of revolution. Only two of these parameters are needed to define the ellipsoid, provided one is linear. As stated earlier, e' is seldom used. Almost universally, the two parameters a and f are used to define the ellipsoid.

Ideally an ellipsoid approximates the shape of the entire earth. During the past few hundred years, there have been dozens of determinations of the earth's form and dimensions. The ones listed in Table 2–1 are most frequently used today. The names of the principal investigators and the year each disclosed his findings, together with the equatorial radius and the flattening, are shown in the table.

TABLE 2–1

Parameters of Commonly Used Ellipsoids

Reference ellipsoid	a (meters)	f
Everest 1830	6,377,304	1/300.8
Bessel 1841	6,377,397	1/299.2
Clarke 1866	6,378,206	1/295.0
Clarke 1880	6,378,249	1/293.5
Hayford 1910	6,378,388	1/297.0
Krasovskii 1938	6,378,245	1/298.3
Hough 1956	6,378,270	1/297.0
Fischer 1960	6,378,166	1/298.3
Kaula 1961	6,378,165	1/298.3
Fischer 1968	6,378,150	1/298.3

A study of Table 2–1 reveals that the difference in equatorial radius between the largest determination (Hayford) and the smallest (Everest) is nearly 1100 meters, which produces a difference in equatorial circumference of about 7000 meters, a sizable quantity. The spread of flattening values in the denominator is over 7. A change of unity in the denominator is equivalent to a difference of approximately 70 meters in the equatorial radius; therefore, the effect of a change of 7 is about 500 meters.

This does not mean that any one of the ellipsoids described in Table 2–1 is necessarily wrong when used as a regional datum (see Chap. 6). A regional datum uses an ellipsoid which best approximates the shape of the earth in that region. This ellipsoid is fixed to the surface at some point in the region. At this surface origin, coordinates and orientations are assumed to be free from error and the survey for the entire region adjusted thereto. For instance, the North American Datum uses the Clarke 1866 ellipsoid with origin at Meade's Ranch in Kansas. The entire continent of North America, parts of South America, and the down-range stations of the Eastern Test Range are all referred to that datum.

Similarly, in other parts of the world, different ellipsoids are used. The Everest ellipsoid is employed in India and Burma. The Bessel ellipsoid was the original one used in Europe and parts of Asia. The Clarke 1880 ellipsoid finds use in Africa. Hayford, the first truly international ellipsoid, is used in most scientific work and also by many countries. Modern Russia has adopted the Krasovskii ellipsoid. The Hough ellipsoid has found favor with the U.S. Army Map Service. The National Aeronautics and Space Administration decided in favor of the Mercury datum, with an ellipsoid derived by Fischer, for use in the Mercury and Gemini programs. The U.S. Department of Defense has its own World Geodetic System known as the DOD WGS.

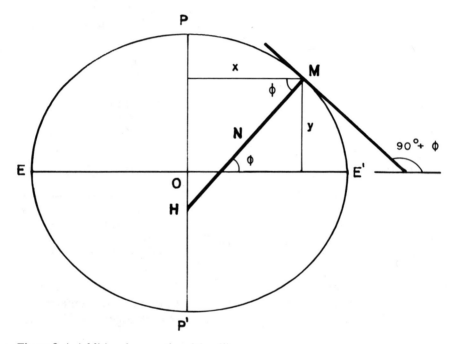

Figure 2–4. Additional properties of the ellipse.

With the principal parameters of the ellipsoid of revolution defined, we continue with its other properties. In Fig. 2–4, a normal to the ellipsoid at point M is drawn until it intersects the minor axis at H. The distance MH is called the normal and is designated N; it is identical to the radius of curvature in the prime vertical through M. This value of N should not be confused with the geoid height shown in Fig. 2–1, which is also called N. There are simply not enough letters in the combined English and Greek alphabets to avoid multiple use in geodesy. The angle made by the normal with the plane of the earth's equator is ϕ, the geodetic (or geographic) latitude. The horizontal distance from the polar axis to any point M on the surface of the ellipsoid is x; the vertical distance from the equatorial plane to the same point is y.

The basic equation of the ellipse is

$$\frac{x^2}{a^2} + \frac{y^2}{b^2} = 1. \tag{2-6}$$

In order to find the coordinates of any point M in terms of latitude it is necessary to differentiate Eq. (2–6), which gives

$$\frac{y}{x} = -\frac{b^2}{a^2}\frac{dx}{dy}. \tag{2-7}$$

The tangent line to an ellipse makes an angle with the horizontal whose tangent is dy/dx, so that

$$\tan(90° + \phi) = \frac{dy}{dx}$$

or

$$\tan \phi = -\frac{dx}{dy}. \tag{2-8}$$

Taking the value of b^2/a^2 from Eq. (2–3) and the value of dx/dy from Eq. (2–8), and substituting these into Eq. (2–7), produces

$$\frac{y}{x} = (1 - e^2) \tan \phi. \tag{2-9}$$

Squaring Eq. (2–9) yields

$$\frac{y^2}{x^2} = (1 - e^2)^2 \tan^2 \phi. \tag{2-10}$$

Rewriting Eq. (2–6) into a different form gives

$$x^2 + \frac{a^2 y^2}{b^2} = a^2.$$

Since

$$\frac{a^2}{b^2} = \frac{1}{1 - e^2},$$

we may write

$$x^2 + \frac{y^2}{1 - e^2} = a^2,$$

or

$$x^2 = a^2 - \frac{y^2}{1 - e^2}; \tag{2-11}$$

then

$$y^2 = (a^2 - x^2)(1 - e^2). \tag{2-12}$$

Substituting the value of y^2 from Eq. (2–12) into Eq. (2–10), we now have

$$\frac{(a^2 - x^2)\,(1 - e^2)}{x^2} = (1 - e^2)^2 \tan^2 \phi,$$

$$\frac{a^2 - x^2}{x^2} = (1 - e^2) \tan^2 \phi,$$

$$a^2 - x^2 = x^2\,(1 - e^2) \tan^2 \phi,$$

or

$$a^2 = x^2 + x^2\,(1 - e^2) \tan^2 \phi;$$

then

$$a^2 = x^2[1 + (1 - e^2) \tan^2 \phi],$$

and

$$x^2 = \frac{a^2}{1 + \tan^2 \phi - e^2 \tan^2 \phi}.$$

But $1 + \tan^2 \phi = \sec^2 \phi$, and $\tan^2 \phi = \sin^2 \phi / \cos^2 \phi$, so

$$x^2 = \frac{a^2}{\sec^2 \phi - e^2 \sin^2 \phi / \cos^2 \phi}.$$

Multiplying through by $\cos^2 \phi$ leaves

$$x^2 = \frac{a^2 \cos^2 \phi}{1 - e^2 \sin^2 \phi},$$

from which

$$x = \frac{a \cos \phi}{(1 - e^2 \sin^2 \phi)^{1/2}}. \tag{2-13}$$

Similarly, if the value of x^2 from Eq. (2–11) is substituted into Eq. (2–10), there results

$$\frac{y^2}{a^2 - y^2/(1 - e^2)} = (1 - e^2)^2 \tan^2 \phi,$$

so that

$$a^2 - \frac{y^2}{1 - e^2} = \frac{y^2}{(1 - e^2)^2 \tan^2 \phi},$$

$$a^2 = \frac{y^2}{1 - e^2} + \frac{y^2}{(1 - e^2)^2 \tan^2 \phi};$$

continuing, we have

$$a^2 = y^2 \left[\frac{1}{1 - e^2} + \frac{1}{(1 - e^2)^2 \tan^2 \phi} \right],$$

$$a^2 = y^2 \left[\frac{(1 - e^2) \tan^2 \phi + 1}{(1 - e^2)^2 \tan^2 \phi} \right],$$

$$a^2 = y^2 \left[\frac{\tan^2 \phi - e^2 \tan^2 \phi + 1}{(1 - e^2)^2 \tan^2 \phi} \right].$$

But $1 + \tan^2 \phi = \sec^2 \phi$, so

$$a^2 = y^2 \left[\frac{\sec^2 \phi - e^2 \tan^2 \phi}{(1 - e^2)^2 \tan^2 \phi} \right],$$

which can be rewritten

$$y^2 = \frac{a^2 (1 - e^2)^2 \tan^2 \phi}{\sec^2 \phi - e^2 \tan^2 \phi},$$

or

$$y^2 = \frac{a^2 (1 - e^2)^2 \sin^2 \phi / \cos^2 \phi}{\sec^2 \phi - e^2 \sin^2 \phi / \cos^2 \phi}.$$

Multiplying through by $\cos^2 \phi$ produces

$$y^2 = \frac{a^2 (1 - e^2)^2 \sin^2 \phi}{1 - e^2 \sin^2 \phi},$$

which finally gives

$$y = \frac{a(1 - e^2) \sin \phi}{(1 - e^2 \sin^2 \phi)^{1/2}}. \tag{2-14}$$

The next two fundamental properties of the ellipsoid are the radius of curvature in the plane of the meridian, designated R, and the radius of curvature in the plane of the prime vertical, known as N. The radius of curvature in the plane of the meridian is quite easy to comprehend. Figure 2–5 is a sketch in the plane of the meridian. It can be seen that the radius of curvature is shorter in the area of the equator than it is in the flatter arc at the pole. Between these two extremes, the value of R increases from the equator to the pole.

To derive the radius of curvature in the meridian, we make use of the general equation of curvature,

$$R = \frac{[1 + (dy/dx)^2]^{3/2}}{d^2y/dx^2}. \tag{2-15}$$

From Eq. (2–7),

$$\frac{dy}{dx} = - \frac{x}{y} \frac{b^2}{a^2}. \tag{2-16}$$

Differentiating, we get

$$\frac{d^2y}{dx^2} = - \frac{b^2}{a^2} \left(\frac{y - x \, dy/dx}{y^2} \right),$$

or

$$\frac{d^2y}{dx^2} = - \frac{b^2}{a^2 y^2} \left(y + \frac{x^2}{y} \cdot \frac{b^2}{a^2} \right).$$

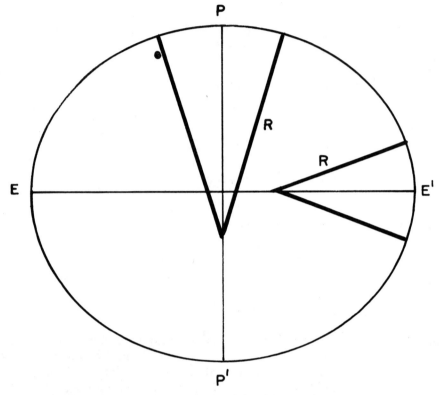

Figure 2–5. Radius of curvature in the plane of the meridian.

Multiplying through by y/b^2 gives

$$\frac{d^2y}{dx^2} = - \frac{b^4}{a^2y^3} \left(\frac{y^2}{b^2} + \frac{x^2}{a^2} \right). \qquad (2\text{–}17)$$

Now, from the basic equation of the ellipse, it can be seen that the portion of Eq. (2–17) which is in parentheses is equal to one, so that we may write

$$\frac{d^2y}{dx^2} = - \frac{b^4}{a^2y^3}. \qquad (2\text{–}18)$$

Substituting the value of dy/dx as given by Eq. (2–16) and the value of d^2y/dx^2 shown in Eq. (2–18) into Eq. (2–15) yields

$$R = - \frac{\left[1 + \dfrac{x^2}{y^2} \dfrac{b^4}{a^4} \right]^{3/2}}{b^4/a^2y^3},$$

$$R = - \frac{[a^4y^2 + b^4x^2]^{3/2}}{a^4b^4}.$$

Then, putting in the values of x and y as given by Eqs. (2–13) and (2–14), we get

$$R = - \frac{\left[\dfrac{a^6(1-e^2)^2 \sin^2 \phi}{1 - e^2 \sin^2 \phi} + \dfrac{b^4 a^2 \cos^2 \phi}{1 - e^2 \sin^2 \phi} \right]^{3/2}}{a^4 b^4}. \tag{2–19}$$

From Eq. (2–3), we know that $b^2 = a^2(1 - e^2)$. By substituting this value of b^2 into Eq. (2–19), we finally have

$$R = - \frac{a(1 - e^2)}{(1 - e^2 \sin^2 \phi)^{3/2}}. \tag{2–20}$$

The negative sign is meaningless and indicates only the direction of bending. The value of R is always regarded as positive.

The value of N, the radius of curvature in the prime vertical, is easier to derive but more difficult to visualize. It is the radius of curvature through any point M on the surface of the ellipsoid in a plane at right angles to the meridian. Many beginning students of geodesy confuse this with the radius of a parallel of latitude, designated R_p. However, the parallel of latitude is not a great circle; it is a small circle, the trace produced by the intersection of the ellipsoid and a plane parallel to the equatorial plane. The parallel encircles the minor axis and intersects every meridian at a right angle. The prime vertical is perpendicular to the meridian at point M, but intersects only one other meridian at a right angle and that is at the antipodal point. Figure 2–6 should help clarify the difference between N and R_p.

The value of N can be derived by reference to Fig. 2–4.

$$N = \frac{x}{\cos \phi}.$$

Into this equation we can substitute the value of x from Eq. (2–13).

$$N = \frac{(a \cos \phi)/(1 - e^2 \sin^2 \phi)^{1/2}}{\cos \phi} = \frac{a}{(1 - e^2 \sin^2 \phi)^{1/2}}. \tag{2–21}$$

The value of R_p, the radius of curvature of a parallel of latitude, can be ascertained quickly by reference to Figs. 2–4 and 2–6. At any point M, it is identical to the value of x.

$$R_p = x = \frac{a \cos \phi}{(1 - e^2 \sin^2 \phi)^{1/2}} = N \cos \phi. \tag{2–22}$$

A thorough understanding of x, y, R, and N is absolutely essential. However, in practice they are seldom computed. Neither x nor y is used extensively except as a way station in obtaining R and N. The values of the radii of curvature R and N have been computed for each minute of

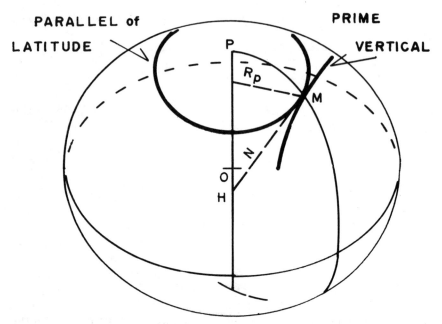

Figure 2–6. Comparison of the parallel of latitude with the prime vertical through a common point.

latitude, together with the difference for each second, from equator to pole, and are available for the commonly accepted ellipsoids [1–7]. It is necessary only to have the appropriate manual for each reference ellipsoid in order to obtain the values of R and N.

The values of R and N as computed by Eqs. (2–20) and (2–21), or as obtained from the manuals of latitude functions, are correct in the meridian and prime vertical, respectively. For any other plane, the true value of the radius of curvature lies somewhere between the two. For any azimuth α measured from north, the radius of curvature is

$$R_\alpha = \frac{RN}{R \sin^2 \alpha + N \cos^2 \alpha}. \qquad (2\text{–}23)$$

This is known as Euler's theorem. To use the theorem, the values of R and N for the specific latitude in question must be computed, or taken from tables, [1–7] and the sine and cosine functions must be obtained for the azimuth of the plane in which the curvature is desired. It is interesting to note that, for the radius of curvature in the plane of the meridian, where $\alpha = 0°$,

$$R_\alpha = \frac{RN}{N} = R,$$

which, of course, is correct.

In the plane of the prime vertical, where $\alpha = 90°$,

$$R_\alpha = \frac{RN}{R} = N,$$

which is the correct value. Between the extremes of $0°$ and $90°$ azimuth, the value of R will be somewhere between R and N.

There are only two points on the ellipsoid where curvature is the same in any azimuth therefrom. These are the poles, P and P'. This statement should be accepted without proof from the very nature of the symmetry of an ellipsoid with respect to its poles. It can, however, be shown mathematically. First, solve for the value of R at the pole by use of Eq. (2–20).

$$R = \frac{a(1 - e^2)}{(1 - e^2 \sin^2 \phi)^{3/2}} = \frac{a(1 - e^2)}{(1 - e^2)^{3/2}} = \frac{a}{(1 - e^2)^{1/2}}.$$

From Eq. (2–3), $(1 - e^2)^{1/2} = b/a$, so

$$R = \frac{a}{b/a} = \frac{a^2}{b}.$$

Similarly, solve for N at the pole by means of Eq. (2–21):

$$N = \frac{a}{(1 - e^2 \sin^2 \phi)^{1/2}} = \frac{a}{(1 - e^2)^{1/2}} = \frac{a}{b/a} = \frac{a^2}{b}.$$

Therefore, the values for R and N are identical at the pole. The radius of curvature at the pole is generally called c, so that

$$c = a^2/b. \tag{2–24}$$

For many computations in geodesy, the geometric mean of R and N is sufficiently accurate. This mean value is designated r and defined as

$$r = (RN)^{1/2}. \tag{2–25}$$

Another property of the ellipsoid often used in geodetic work is the meridional arc, often abbreviated MA. It is defined as the true meridional distance along the surface of the ellipsoid measured from the equator to any point M. The length of an arc of the meridian between any two points M_1 and M_2 is s. Figure 2–7 shows the geometry. The value of s is given by

$$s = a(1 - e^2) \int_{\phi_1}^{\phi_2} \left(1 + \frac{3}{2} e^2 \sin^2 \phi + \frac{15}{8} e^4 \sin^4 \phi + \frac{35}{16} e^6 \sin^6 \phi + \cdots \right) d\phi. \tag{2–26}$$

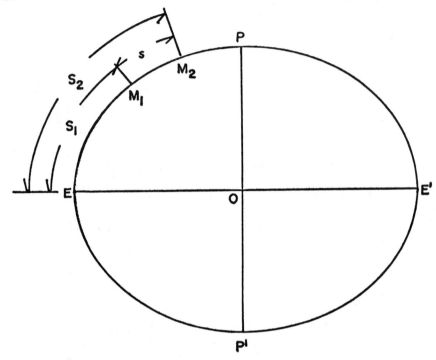

Figure 2–7. The meridional arc.

This would be a tedious problem to solve if the tables of latitude functions did not provide a ready solution. The meridional arc is given for each minute, with differences for each second of latitude. This is, of course, the distance from the equator. To obtain the distance s along the meridian between any two points M_1 and M_2, simply extract from the tables the meridional arcs S_1 and S_2 corresponding to points M_1 and M_2, respectively; then find the difference,

$$s = S_2 - S_1. \qquad (2\text{--}27)$$

TYPES OF LATITUDE

In his study of geometrical geodesy, the reader will encounter a number of different types of latitude. Only three are of practical importance. These are the geodetic (also known as geographic) latitude, the geocentric latitude, and the reduced latitude. The three are defined by reference to Fig. 2–8.

From any point M on the surface of a reference ellipsoid, a normal drawn to the reference ellipsoid will intersect the equatorial plane EE'. The angle between the normal and the equatorial plane is the geodetic latitude and is designated ϕ.

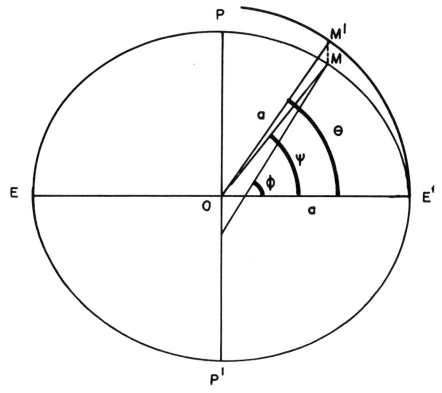

Figure 2–8. Three types of latitude.

From the same point M, a radius vector may be drawn to O, the center of the ellipse. The angle between the line MO and the equatorial plane is the geocentric latitude and is generally designated ψ.

Through the same point M, extend the y coordinate until it intersects an auxiliary circle, known as the major circle of the ellipse, having its center at the center of the ellipse and a radius equal to the semimajor axis. This new point is M'. The angle between the radius vector $M'O$ and the equatorial plane is the reduced latitude, or eccentric angle, and is generally designated θ.

The relationships among these three types of latitude are as follows:

$$\tan \psi \;=\; (1 - e^2) \quad \tan \phi \;=\; \frac{b^2}{a^2} \; \tan \phi, \qquad (2\text{–}28a)$$

$$\tan \psi \;=\; (1 - e^2)^{1/2} \tan \theta \;=\; \frac{b}{a} \; \tan \theta, \qquad (2\text{–}28b)$$

$$\tan \theta = (1 - e^2)^{1/2} \tan \phi = \frac{b}{a} \tan \phi, \qquad (2\text{--}28\text{c})$$

$$\tan \theta = \frac{\tan \psi}{(1 - e^2)^{1/2}} = \frac{a}{b} \tan \psi, \qquad (2\text{--}28\text{d})$$

$$\tan \phi = \frac{\tan \psi}{1 - e^2} = \frac{a^2}{b^2} \tan \psi, \qquad (2\text{--}28\text{e})$$

$$\tan \phi = \frac{\tan \theta}{(1 - e^2)^{1/2}} = \frac{a}{b} \tan \theta. \qquad (2\text{--}28\text{f})$$

A study of these equations and Fig. 2–8 reveals that $\phi > \theta > \psi$.

Very often there is a need to convert from one type of latitude to another. Most generally the problem is one of converting from geodetic to geocentric or vice versa. The following equation can be used in place of the relationships shown above:

$$\phi - \psi = \tfrac{1}{2}\rho e^2 \sin 2\phi. \qquad (2\text{--}29)$$

The term e^2 has been explained before. The term ρ is merely the conversion of radian measurement into degrees, minutes, or seconds of arc, as follows:

$$1 \text{ radian} = 57.29578° = 3437.747' = 206{,}264.8''.$$

The constants in Eq. (2–29) may be consolidated depending on the ellipsoid to be used and the choice of angular units For the Clarke 1866 ellipsoid, using the minute as the angular unit, we have

$$\phi - \psi = \tfrac{1}{2} \cdot 3437.747 \cdot 0.00676\ 86579\ 97291 \sin 2\phi = 11.634 \sin 2\phi. \qquad (2\text{--}30)$$

From the equation, it is apparent that the difference between ϕ and ψ is zero at latitudes of 0° and 90°, and reaches a maximum when the latitude is 45°.

One other type of latitude will be discussed later in geodetic astronomy. It is astronomical latitude and is defined as the angle between the direction of gravity (plumbline) and the equatorial plane. The astronomical latitude is called Φ to distinguish it from geodetic latitude.

THE TRIAXIAL ELLIPSOID

The triaxial ellipsoid was mentioned briefly earlier in the chapter. While it is ruled out as a mathematical surface because of the complexity of computations, it is still of more than casual interest. Since 1859, there have been many determinations of the best-fitting triaxial ellipsoid.

Figure 2–9 explains the parameters used to describe the triaxial ellipsoid. The polar semiminor axis is known as b, as in a standard ellipsoid. The equator is an ellipse with semimajor axis a_1 and semiminor axis a_2. Table 2–2 lists the various investigators, together with the value of a_1 minus a_2 and the longitude λ_1 of the major equatorial axis [8]. If the findings of all investigators are averaged, which is admittedly not a scientific approach, the value of a_1 minus a_2 is slightly less than 500 meters, with the major axis lying at $1\frac{1}{2}°$ west longitude.

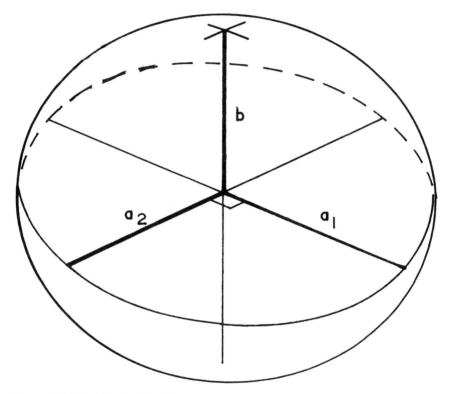

Figure 2–9. The triaxial ellipsoid.

The investigators listed in Table 2–2 have followed three methods: astrogeodetic, gravity, and satellite. The conclusion that can be reached from these data is that a strong case exists in support of the triaxial ellipsoid. In spite of its interest as a scientific and academic problem, the triaxial ellipsoid is not used extensively in geodetic circles. The ellipsoid of revolution as explained in detail in this chapter is the fundamental mathematical surface of geodesy.

TABLE 2–2

Some Triaxial Ellipsoids

Year	Investigator	$a_1 - a_2$ (meters)	λ_1 (degrees)	Method
1859	Schubert	719	41 E	Astrogeodetic
1860	Clarke	1618	14 E	Astrogeodetic
1866	Clarke	1944	15 E	Astrogeodetic
1878	Clarke	465	8 W	Astrogeodetic
1884	Hill	1260	17 E	Astrogeodetic
1915	Helmert	230	17 W	Gravimetric
1916	Berroth	150	10 W	Gravimetric
1924	Heiskanen	345	18 E	Gravimetric
1928	Heiskanen	242	0	Gravimetric
1929	Heiskanen	165	38 E	Astrogeodetic
1934	Hirvonen	139	19 W	Gravimetric
1938	Heiskanen	352	25 W	Gravimetric
1942	Krasovskii	213	15 E	Astrogeodetic
1945	Lambert	352	25 W	Gravimetric
1945	Niskanen	293	4 W	Gravimetric
1957	Uotila	135	6 W	Gravimetric
1961	Izsak	205	33 W	Satellite
1961	Kozai	148	37 W	Satellite

REFERENCES

1. Department of the Army, Latitude Functions, Hayford (International) Ellipsoid, *Technical Manual 5–241–17*, Government Printing Office, Washington, D.C.
2. Department of the Army, Latitude Functions, Clarke 1866 Ellipsoid, *Tech. Manual 5–241–18*, Government Printing Office, Washington, D.C.
3. Department of the Army, Latitude Functions, Bessel Ellipsoid, *Tech. Manual 5–241–19*, Government Printing Office, Washington, D.C.
4. Department of the Army, Latitude Functions, Clarke 1880 Ellipsoid, *Tech. Manual 5–241–20*, Government Printing Office, Washington, D.C.
5. Department of the Army, Latitude Functions, Everest Ellipsoid, *Tech. Manual 5–241–21*, Government Printing Office, Washington, D.C.
6. Department of the Army, Latitude Functions, Fischer 1960 Ellipsoid, *Tech. Manual 5–241–35*, Government Printing Office, Washington, D.C.
7. Department of Defense, Latitude Functions, Krasovskii Ellipsoid, *Geodetic Library file number B–910.6117*.
8. W. A. Heiskanen, Is the earth a triaxial ellipsoid?, *J. Geophys. Res.*, **67**, No. 1 (1962).

Chapter 3 Calculation of Triangulation

In order to cover large areas of the earth's surface, such as an entire continent, with a network of geodetic positions, it is necessary to have some system for extending these positions from some point of origin. These widely scattered positions, where latitude, longitude, elevation, and azimuth have been accurately determined, are then used to control surveys for such purposes as mapping and charting, determining national and state boundaries, railroad and highway construction, and missile site location. For the more limited areas, plane rectangular coordinates are sometimes used; however, for large areas, as in the case of a country or continent, the latitude and longitude system is highly preferred.

The means for providing this widespread system of coordinates is a network of triangles with vertices that are permanently marked on the earth's surface. These points are known as triangulation stations. The length of one side of some triangle in the system must be measured directly. This is called the baseline, or simply the base. Additionally the coordinates of one of the triangulation stations must be known, as an initial point, as well as the azimuth of one of the lines. This information provides a starting point, a direction, and a distance. A mathematical surface, one of the reference ellipsoids described in the previous chapter, must be used by custom, assumption, or direction. With these data already known, the horizontal angles in the network of triangles must now be measured in order to allow calculation of the remaining triangle sides. In addition to these measurements mandatory for calculations, it is highly desirable to have other measurements for the purpose of checking the accuracy of both the field work and the calculations. These additional measurements may be baselines, astronomical azimuths, or check angles. These redundant observations allow the entire triangulation to be "adjusted," by methods which will be described in Chapter 11, to remove inconsistencies from the computed results.

Traditionally, the polygon with a central point station and the completed quadrilateral are the two geometric figures most universally used in triangulation. The polygon with an interior station is an extremely strong figure and is particularly useful in surveys of areas having nearly equal dimensions. Many of the European countries use this geometric figure. For a country of vast dimensions, such as the United States, triangulation belts are used and the best geometric figure is the completed quadrilateral. This figure consists of four stations in the form of a quadrilateral, with its two diagonals forming four triangles in

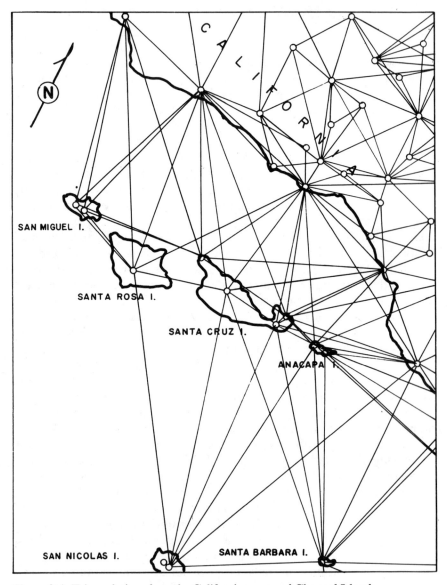

Figure 3-1. Triangulation along the California coast and Channel Islands.

which eight independent angles can be measured. The U.S. Coast and Geodetic Survey has employed the quadrilateral as the basic geometric figure in its principal triangulation belts. Figure 3–1 shows a portion of a triangulation belt in the United States.

Triangulation systems have been classified in many ways depending upon the time, the place, and the agency. One general classification has been with reference to its use and its relationship with other triangulation in the same system. Classified in this manner, it has been called main scheme, basic, principal, and subsidiary. Another general classification has been according to the degree of accuracy obtained. Around this classification have sprung such terms as primary, precise, first order, secondary, second order, tertiary, third order, and fourth order. These descriptive terms refer to criteria which are established by the agency doing the classifying. The best procedure is to obtain a current dictionary of surveying and mapping terms published by the leading geodetic agency of the country in which operations are conducted.

SPHERICAL EXCESS

Essentially, the difference between plane surveying and geodetic surveying is the difference between a plane triangle and a spherical triangle. One of the first facts which the plane surveyor-turned-geodesist learns about geodetic triangulation is that the three interior angles of a spherical triangle do not add up to 180°, but to a value in excess of 180°. As an illustration, let us consider a large spherical triangle on the earth's surface. Let one apex fall at the intersection of the meridian of Greenwich and the equator; this angle is 90°. Assume that the second angle is at the intersection of the 90th meridian west longitude and the equator; this angle is 90° also. Here we have only two of the angles of the spherical triangle measured, yet the total is already 180°. Now add the third angle, which is the intersection of the Greenwich meridian and the 90th meridian. This polar angle is 90°. Adding these three angles produces 270°, which is 90° in excess of what the plane surveyor has always used in his plane triangle. Of course, this spherical triangle is larger than those used in a triangulation network; however, it is illustrative of the fact that the three angles of a spherical triangle exceed 180° by an amount which is known as the spherical excess and designated ϵ. In Fig. 3–2, which is a spherical triangle with angles α, β, and γ and sides a, b, and c, the following equation expresses the sum of the interior angles:

$$\alpha + \beta + \gamma = 180° + \epsilon. \tag{3–1}$$

From solid geometry, the spherical excess of a spherical triangle is exactly proportional to the area of the triangle. In the case of a spheroidal triangle, which is the kind found on an ellipsoid of revolution, the

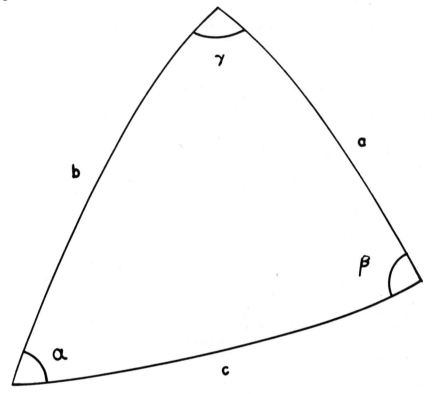

Figure 3–2. Spherical triangle.

same ratio is approximately true, and, for all practical geodetic purposes, can be considered exact.

Since spherical excess is proportional to the area of the spherical
triangle, it must be expressed in terms of the dimensions of the triangle.
Referring to Fig. 3–2, we can express the spherical excess in terms of the
length of two sides and the size of the included angle, or

$$\epsilon = \frac{\rho}{2RN} ab \sin \gamma = \frac{\rho}{2RN} bc \sin \alpha = \frac{\rho}{2RN} ac \sin \beta, \qquad (3\text{–}2)$$

where ϵ is the spherical excess in seconds of arc; ρ is 206,264.8, the number of seconds of arc in a radian; R is the radius of curvature in the
meridian; N is the radius of curvature in the prime vertical; α, β, and
γ are the angles of the spherical triangle; and a, b, and c are the sides of
the spherical triangle.

In practice, the quantity $\rho/2RN$ is called m and has been computed
and listed for each half-degree of latitude in each of the tables of
latitude functions referred to in Chapter 2. The latitude to be used in
obtaining m from the tables is the mean latitude of the three vertices of

the triangle. It is sufficiently accurate to accept the value of m for the nearest half-degree listing; interpolation is not necessary.

By substitution of m for $\rho/2RN$, Eq. (3–2) can be rewritten,

$$\epsilon = mab \sin \gamma = mbc \sin \alpha = mac \sin B, \qquad (3\text{–}3)$$

where a, b, and c are expressed in meters and ϵ in seconds of arc.

For triangles having sides over 100 miles in length, the following more accurate equation should be used:

$$\epsilon_1 = \epsilon + \epsilon \left(\frac{a^2 + b^2 + c^2}{24RN} \right) = \epsilon \left(1 + \frac{a^2 + b^2 + c^2}{24RN} \right), \qquad (3\text{–}4)$$

where ϵ is the spherical excess in seconds of arc as determined in the first iteration by means of Eq. (3–3); ϵ_1 is the final refined value in seconds of arc; and a, b, c, R, and N are as previously defined.

THE AUXILIARY PLANE TRIANGLE

For ordinary triangles on the earth's surface, the error in calculating spheroidal triangles as spherical triangles is negligible. For this reason, the use of spherical triangles is possible and the solution of a single spherical triangle is not difficult. The equations of spherical trigonometry are not in themselves unduly complicated. Appendix A lists the more commonly used equations of spherical trigonometry. However, the direct solution of a series of spherical triangles in an extended survey net would be extremely complicated. On the other hand, the solution of a series of plane triangles is quite simple. Only the plane law of sines is used. It is stated as follows:

In any plane triangle, the ratio of any two sides is equal to the ratio of the sines of the angles opposite these sides in the same order.

Stated mathematically, with reference to Fig. 3–3, the law of sines is

$$\frac{a}{b} = \frac{\sin A}{\sin B},$$

$$\frac{b}{c} = \frac{\sin B}{\sin C},$$

$$\frac{a}{c} = \frac{\sin A}{\sin C}.$$

Written in a different form, the law of sines is

$$\frac{a}{\sin A} = \frac{b}{\sin B} = \frac{c}{\sin C}.$$

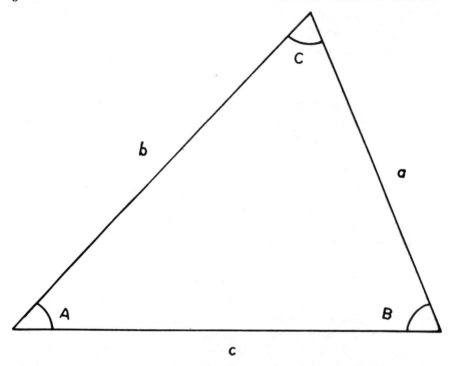

Figure 3–3. Plane triangle.

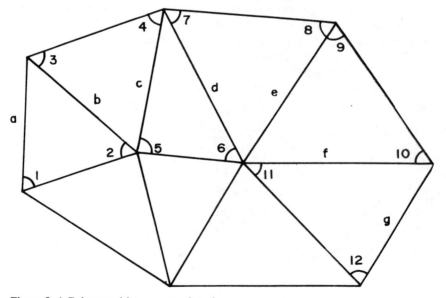

Figure 3–4. Polygon with two central stations.

The following example illustrates the simplicity of dealing with plane triangles in a network. Figure 3–4 is a polygon with two central stations. It is desired that we find the length of line g by use of a chain of triangles which have been propagated from baseline a. By the plane law of sines, and working toward line g from baseline a, we can write in turn

$$b = a \frac{\sin 1}{\sin 2},$$

$$c = b \frac{\sin 3}{\sin 4},$$

$$d = c \frac{\sin 5}{\sin 6},$$

$$e = d \frac{\sin 7}{\sin 8},$$

$$f = e \frac{\sin 9}{\sin 10},$$

$$g = f \frac{\sin 11}{\sin 12}.$$

Then, by substitution of the value of each preceding equation into the final equation, we may write

$$g = a \frac{\sin 1 \sin 3 \sin 5 \sin 7 \sin 9 \sin 11}{\sin 2 \sin 4 \sin 6 \sin 8 \sin 10 \sin 12}.$$

By systematic numbering of the angles, the equation can be written directly by inspection. The same scheme could have been followed had we chosen to solve for the value of g by going through the lower chain of triangles, or if we had wished to compute completely around the net and solve for the value of side a, as a check on the base. Imagine the complications inherent in the use of spherical trigonometry!

In order to avoid these complications, we need an auxiliary plane triangle which can be substituted for its corresponding spherical triangle but which will still provide the same results. We need a method which will give us the mathematical authority for this simplification. As an answer to this dilemma, two solutions were derived, one known as Legendre's theorem and the other called the method of addends.

LEGENDRE'S THEOREM
The use of spherical triangles may be avoided entirely by employing a principle known as Legendre's theorem. A full statement of the theorem is

If we have a spherical triangle whose sides are short compared with the radius of the sphere, and if we also have a plane triangle whose sides are equal in length to the corresponding sides of the spherical triangle, then the corresponding angles of the two triangles differ by approximately the same quantity, which is one-third the spherical excess of the triangle.

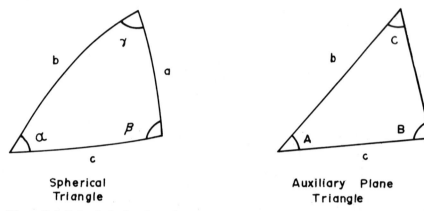

Spherical
Triangle

Auxiliary Plane
Triangle

Figure 3–5. Spherical triangle and auxiliary plane triangle (Legendre's method).

This principle allows us to use an auxiliary plane triangle instead of the spherical triangle. In Fig. 3–5, a spherical triangle is shown on the left, and the corresponding plane triangle on the right. By Legendre's theorem, the corresponding sides are equal in length but the corresponding angles differ by one third of the spherical excess of the triangle. Stated mathematically, the plane angles are

$$A = \alpha - \frac{\epsilon}{3},$$ (3–5a)

$$B = \beta - \frac{\epsilon}{3},$$ (3–5b)

$$C = \gamma - \frac{\epsilon}{3}.$$ (3–5c)

After the field observations have been completed, the spherical excess is computed by means of either Eq. (3–3) or (3–4), depending on the size of the triangle. One third of this spherical excess is subtracted from each angle of the spherical triangle to obtain the corresponding plane angle. If there were no errors in the observations, the sum of the plane angles would be exactly 180°. The difference between the sum of the plane angles and 180° is the error of observation. This error may be distributed equally among the three angles unless weights are assigned to the observations of each of the three angles. In a triangulation network, where we are dealing with a multiple number of triangles, weights are generally assigned to the observations and a least-square adjustment is made. Once the error is distributed so that the sum of the three angles is forced to equal 180°, the lengths of the two remaining sides of the triangle may be found by the law of sines from plane trigonometry.

It must be stressed that the plane angles are to be used *only* in the solution of the triangle. Once the sides of the triangle have been computed, the spherical angles must be used when calculating the geographic coordinates of the stations.

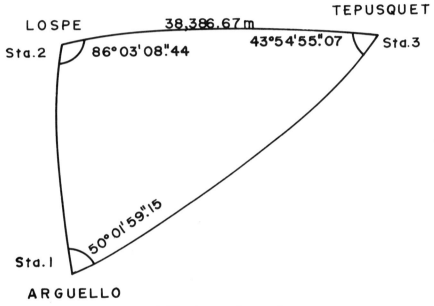

Figure 3–6. Single triangle in California coastal network.

An example of the foregoing procedure may be helpful. Figure 3–6 is a single triangle with apices at the three triangulation stations Arguello, Lospe, and Tepusquet. This is one of the triangles in the network covering Vandenberg Air Force Base in California and is a real-life example from the files of the U.S. Coast and Geodetic Survey. All the angles have been observed and the side Lospe–Tepusquet previously determined by adjacent triangulation. A tabular arrangement is convenient for this problem, as shown by Table 3–1.

First, the observed angles are added and found to be in excess of 180° by the amount of 2.″66. The spherical excess is then computed by

TABLE 3-1

Tabular Arrangement of Triangle Calculation

Stations	Observed angles	Correc- tions	Spherical angles	Sphe- rical excess	Plane angles	Distance (meters)
2–3						38,386.67
1 Arguello	50° 01′ 59″.15	+0″.24	50° 01′ 59″.39	1″.13	50° 01′ 58″.26	
2 Lospe	86 03 08.44	+0.24	86 03 08.68	1.12	86 03 07.56	
3 Tepusquet	43 54 55.07	+0.24	43 54 55.31	1.13	43 54 54.18	
	180 00 02.66	+0.72	180 00 03.38	3.38	180 00 00.00	
1–3						49,967.30
1–2						34,739.31

Eq. (3–3) and is 3″.38. Only one side of the triangle is known, while Eq. (3–3) demands that two sides be known. In the calculation of spherical excess, it is sufficiently accurate to scale the length of the sides from a large scale map, or to make a rough determination of the length by the best *a priori* information available.

If there had been no errors in observation, the observed angles would have added to 180° 00′ 03″.38. This means that 0″.72 (the difference between 3″.38 and 2″.66) must be attributed to errors in observation. One third of this error is arbitrarily applied to each observed angle to get the corrected spherical angle. One third of the spherical excess is subtracted from each corrected spherical angle to obtain the plane angles, which then total 180°. By plane trigonometry, using the law of sines, the distances Arguello–Tepusquet and Arguello–Lospe are computed. Once the plane triangle is solved for these two distances, it is necessary to revert to the spherical angles to compute the geographic coordinates of the unknown stations.

While this example is for a single triangle only, the procedure is similar when one is considering multiple triangles in a survey net.

THE METHOD OF ADDENDS

Another method is available for solving the spherical triangles of a survey net by use of auxiliary plane triangles. This is the "method of additament," a term which comes from the German *additamenten-methode*. More simply, it is called the method of addends and was introduced in Bavaria by Soldner at the beginning of the past century. It is used very little in the United States but has found some favor in Europe.

In Legendre's theorem, we converted from spherical to plane triangles by subtracting one third of the spherical excess from the corrected observed angles. In the method of addends, the corrected observed angles (spherical angles) are used as they are, but the sides are shortened. The following statements make clear the essential difference between the two.

Legendre: The sides of the triangle are held fixed while the angles are reduced in size.

Addends: The angles are held fixed while the sides are shortened.

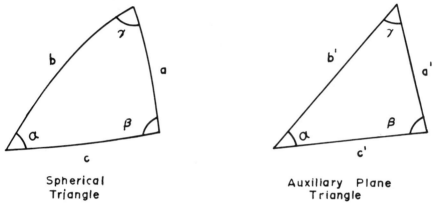

Spherical
Triangle

Auxiliary Plane
Triangle

Figure 3–7. Spherical triangle and auxiliary plane triangle (method of addends).

Figure 3–7 shows a spherical triangle and its auxiliary plane triangle. The angles are the same but the sides a', b', and c' in the plane triangle are shorter than their corresponding sides a, b, and c in the spherical triangle. The shortened lengths to be used are

$$a' = a\left(1 - \frac{a^2}{6r^2} \cdots \right), \tag{3–6a}$$

$$b' = b\left(1 - \frac{b^2}{6r^2} \cdots \right), \tag{3–6b}$$

$$c' = c\left(1 - \frac{c^2}{6r^2} \cdots \right); \tag{3–6c}$$

or, in general terms,

$$s' = s\left(1 - \frac{s^2}{6r^2} \cdots \right). \tag{3–7}$$

In logarithmic form, the general expression is

$$\log s' = \log s - M\frac{s^2}{6r^2} \cdots, \tag{3–8}$$

where $M = \log_{10} e = 0.4342945$, the modulus of the common logarithms.

In these equations, the shortened sides have been designated by primes. The value of r^2 may be replaced by RN, as indicated by Eq. (2–25). From the form of the equations, it may be noticed that a series expansion has been used. Only the significant terms have been included.

After the base of the first triangle has been reduced by subtracting the correction, the computation of the whole chain of triangles may be carried out, using the spherical angles only. It is not necessary to add the corrections to the computed sides until their true values are needed later for determining the geographic coordinates of the stations.

Either Legendre's theorem or the method of addends can be used, depending on personal preference or established procedures. They give similar results. Either provides the simplification needed to avoid the direct computation of a network of spherical triangles.

Chapter 4 Computations on the Ellipsoid

In Chapter 2, we learned some of the principal properties of the fundamental mathematical surface of the earth, the ellipsoid of revolution.
One of the reference ellipsoids must be used, by custom, assumption, or
direction. The Clarke 1866 ellipsoid is customarily used in the United
States. Elsewhere in the world, the geodesist must use the reference
ellipsoid of that particular country or region. First and foremost, we
need a mathematical surface upon which to compute.

In Chapter 3, we saw how geodetic positions are propagated from a
point of origin through a principal network so that the area is covered
with readily available stations where latitude, longitude, elevation, and
azimuth have been determined. From these survey stations, geodesists
may do subsidiary triangulation or run traverses in order to accomplish some useful function. Methods were presented whereby plane
auxiliary triangles may be used in lieu of spherical triangles so that the
tedium of complicated calculations may be avoided.

In this chapter, we turn our attention to the procedures for making
computations on the ellipsoid for the purpose of obtaining such
meaningful information as length of lines, azimuth, and coordinates.
As the authors stated in their prefacing remarks, this is not a field
manual. The process of making field measurements and their corrections is an art unto itself and is outside the scope of this book. For this
reason, the computations considered in this chapter are made after the
field work is completed.

REDUCTION OF BASELINE TO THE REFERENCE ELLIPSOID

Let us assume that the geodetic surveyor has measured a baseline for
subsequent use in the calculation of triangulation. If he accomplished
his measurements by use of an invar tape, he would have made the
customary corrections for grade, alignment, temperature, change in
weight or position of thermometers, sag, absolute length of tape, and
any other corrections specified by his survey agency. If the measurements were made by electronic methods, such as the Geodimeter, the
Tellurometer, or the Electrotape, the customary corrections of reduction to the horizontal and variation from the accepted velocity of light

would have been performed. This measured and corrected baseline is still not usable in geodetic computations because it lies on the physical surface of the earth. It must be reduced to the surface of the reference ellipsoid and all subsequent computations performed on that mathematical surface.

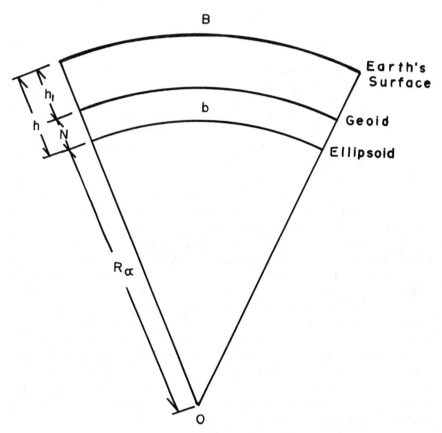

Figure 4–1. Reduction of a baseline to the reference ellipsoid.

In Figure 4–1, B is the baseline as measured and corrected on the surface of the earth. If the two ends of the baseline are at different elevations, the mean value should be used in the reduction to the reference ellipsoid. The baseline we need is b, lying on the surface of the reference ellipsoid. It must be emphasized again that the geoid and the ellipsoid are not the same surfaces. This was mentioned briefly in the introduction and will be explained in detail in Chapter 8. The geoid, which is the sea-level surface projected inland, must be used if there is no knowledge of the location of the reference ellipsoid. Where a sufficiently detailed survey has been accomplished and the geoid height determined, the geoid will have been defined. For purposes of illustration, let us assume that the geoid height is known and that it lies a distance of N above the ellipsoid. By means of leveling, the height h_1 of the earth above sea level has been determined. The sum N and h_1 is indicated by h, which is the distance of the surface from the reference ellipsoid. The notation R_α is the same as explained in Chapter 2 and its value is as shown by Eq. (2–23). It is the radius of curvature of the earth in the plane of the baseline.

By simple proportion,

$$\frac{b}{B} = \frac{R_\alpha}{R_\alpha + h},$$

which can be rewritten

$$b = B \frac{R_\alpha}{R_\alpha + h} = B \frac{1}{1 + h/R_\alpha}. \tag{4-1}$$

By use of the series

$$\frac{1}{1 + x} = 1 - x + x^2 - x^3 + \cdots,$$

we obtain an expansion of Eq. (4–1),

$$b = B \left(1 - \frac{h}{R_\alpha} + \frac{h^2}{R_\alpha{}^2} - \cdots \right); \tag{4-2}$$

or, if the amount of the correction is needed, there results

$$B - b = B \frac{h}{R_\alpha} - B \frac{h^2}{R_\alpha{}^2} + \cdots. \tag{4-3}$$

The geoid height N may be positive or negative depending on whether the geoid is above or below the ellipsoid. Furthermore, the measured baseline B could possibly be below the geoid and the ellipsoid as in the case of a baseline measured in Death Valley or near the Salton Sea. For these reasons, it is suggested that a sketch be drawn in order to place the geoid, the ellipsoid, and the physical surface in their correct relative locations. By reference to the sketch and by the proper use of algebraic signs, the right correction can be determined. Remember that the only simple geometric correction we wish to make is to reduce a length measured on the physical surface to its corresponding length on the ellipsoid.

EFFECT OF HEIGHT OF POINT OBSERVED ON THE AZIMUTH OF A LINE

Because of the flattening of the reference ellipsoid, the normals projected vertically upward from two points on the ellipsoid generally do not lie in the same plane. One of the normals will be inclined with respect to the other. For this reason, there will be an error in the observed azimuth of a station depending on its height above the surface of the ellipsoid. An analogy is found in plane surveying when the range pole is not held precisely in a vertical position above the station. If the sight is made on the base of the pole, there is no error in the horizontal angle. However, there is an error if the sight is taken somewhere up the inclined rod; the higher the point of sighting, the greater the error.

Unless the point sighted upon is at a high elevation (in excess of 1000 meters), the error is smaller than the probable error of an observed direction and, therefore, is negligible. In mountainous terrain where observations may be made upon stations located at extremely high altitudes, as in the western part of the United States, a correction must be applied to the observed direction.

The correction to the azimuth is

$$\Delta\alpha = \frac{\rho h e^2 \cos^2 \phi \sin 2\alpha}{2\mathcal{N}(1 - e^2)}, \tag{4-4}$$

where $\Delta\alpha$ is the correction in seconds of arc; ρ is 206,264.8, the number of seconds in a radian; h is the height of the observed station above the ellipsoid in meters; e is the flattening of the ellipsoid; ϕ is the latitude of the station observed; α is the azimuth to the station observed; and $\mathcal{N}$ is the radius of curvature in the prime vertical at latitude ϕ.

The correction is to be applied as follows:

When the station being sighted upon is northeast or southwest of the observer, the azimuth must be increased to obtain the correct azimuth at the surface of the ellipsoid; when the station being sighted upon is northwest or southeast of the observer, the azimuth must be decreased.

CONVERGENCE OF THE MERIDIANS

At the equator, all meridians are parallel because they intersect the equator at right angles. At the poles, all meridians have converged to a common point. Between the equator and the pole, the amount of convergence can be determined mathematically.

For convenience, let Fig. 4–2 be a perfect sphere rather than an ellipsoid. The figure represents a wedge-shaped slice of width $\Delta\lambda$ bounded by the equator and two meridional planes in the northern hemisphere. The arc AB is the segment of a parallel of latitude lying at latitude ϕ. The radius of the sphere is r. The radius of the parallel of latitude is r cos ϕ. Tangent to the sphere at the points A and B, two lines are extended, each in the plane of its respective meridian. These two lines will intersect along the extension of the polar axis and form angle

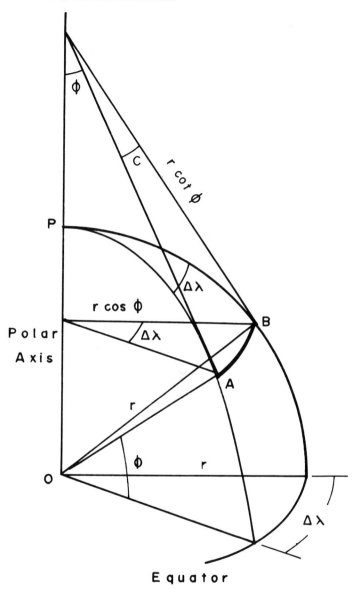

Figure 4–2. Convergence of the meridians.

C, which is the amount of the convergence of the two meridians. The length of the two lines is $r \cot \phi$.

From the figure, we see that AB is common to two triangles, or

$$AB = r \cos \phi \cdot \Delta\lambda \qquad \text{and} \qquad AB = r \cot \phi \cdot C.$$

Equating these two values of AB, we may write

$$r \cos \phi \cdot \Delta\lambda = r \cot \phi \cdot C.$$

from which

$$C = \Delta\lambda \, \frac{\cos \phi}{\cot \phi} = \Delta\lambda \sin \phi. \tag{4-5}$$

This is the amount of convergence at latitude ϕ on a sphere. Equation (4–5) is used for surveys of limited extent, as in the establishment of township and section lines, or when checking the azimuths of lines in a traverse.

The equation for convergence of meridians on an ellipsoid is somewhat more complicated. Its value is

$$C = \Delta\lambda \sin \phi_m \sec \frac{\Delta\phi}{2} + (\Delta\lambda)^3 \cdot F, \tag{4-6}$$

where C is the convergence in seconds of arc; $\Delta\lambda$ is the difference in longitude between points A and B, expressed in seconds of arc; ϕ_m is the average latitude between A and B; $\Delta\phi$ is the difference in latitude between A and B; and F is an abbreviation for $\frac{1}{12} \sin \phi_m \cos^2 \phi_m \sin^2 1''$.

The value of F is given in each manual of latitude functions referred to in Chapter 2. Since F is dependent on latitude only, it does not vary among the various reference ellipsoids.

As a check on the validity of Eq. (4–6), we may substitute values into the equation for the case at the equator and also at the pole. At the equator, the convergence should be zero.

$$C \text{ (at equator)} = \Delta\lambda \cdot 0 \cdot 1 + (\Delta\lambda)^3 \cdot 0 = 0.$$

At the pole, the convergence should be equal to the difference in longitude.

$$C \text{ (at the pole)} = \Delta\lambda \cdot 1 \cdot 1 + (\Delta\lambda)^3 \cdot 0 = \Delta\lambda.$$

The practical use of convergence in geodesy lies in the fact that the forward and back azimuths of a line do not differ by 180°, as is the case in plane surveying, but by 180° plus the convergence.

Figure 4–3 shows two meridians λ_1 and λ_2 converging toward the pole. Each lies in a true north-south line but they are not parallel. Point A lies on the meridian of λ_1 and point B is on the meridian of λ_2. The azimuth of line AB measured with respect to the meridian λ_1 is designated α_{AB}. The azimuth of the line BA with reference to its meridian λ_2 is designated α_{BA}. Through point A, a line is drawn parallel to the meridian λ_2. The angle this line makes with the meridian λ_1 is the convergence C. We may now write

$$\alpha_{BA} = \alpha_{AB} + C + 180°, \tag{4-7}$$

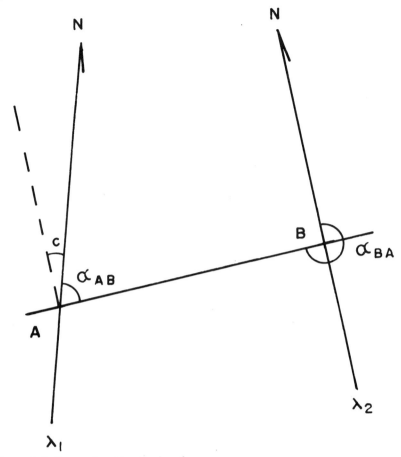

Figure 4–3. Forward and back azimuths.

which shows that the forward and back azimuths of a line differ by 180°
plus the convergence.

THE PLANE CURVES AND THE GEODESIC LINE

On a sphere, the shortest distance between two points A and B is the
arc of a great circle connecting these points. If a theodolite were set up
at point A, leveled, and sighted at point B, the sight plane would trace
out the arc of a great circle. If the instrument were moved to point B,
leveled, and sighted at point A, the sight plane would again trace out
the arc of a great circle. These two arcs would be identical because the
normals at the two points intersect at a common point, the center of the
sphere.

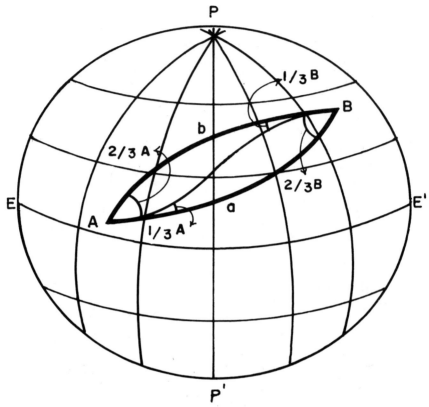

Figure 4–4. The plane curves and the geodesic line.

This is not the case on an ellipsoid. Because of flattening, the normals at different latitudes never intersect. The higher the latitude, the lower the point where the normal intersects the polar axis. If a theodolite is set up at point *A* in Fig. 4–4 and leveled, its vertical axis will coincide with the normal to the ellipsoid at that point. In this discussion, we assume that no gravity anomalies exist and that the deflection of the vertical is zero. If the theodolite is sighted at point *B*, the plane of sight will trace out the line *AaB*. If the theodolite is moved to point *B* and leveled, its vertical axis will now coincide with the normal at that point. Since *B* is at a higher latitude, the normal will intersect the polar axis at a lower point and so the plane of sight from *B* to *A* will trace out the line *BbA*, which is north of the other line. Both planes contain the chord *AB*, but the two normals are tilted with respect to each other so the trace on the surface must be different in each case. These two curves, *AaB* and *BbA*, are called plane curves.

On an ellipsoid, the shortest line that can be drawn between two points is a geodesic line. It is synonymous with the term geodetic line, or simply the geodesic. Unlike the plane curves, the geodesic line has a double curvature. If we were to imagine that the theodolite is moved between A and B of Fig. 4–4 so that the vertical plane of the instrument always contained the normal to the ellipsoid, the path of the theodolite would be the geodesic line. When traced completely around the ellipsoid, the geodesic line will not generally return to the starting point but will pass that point on the equator in a slightly different longitude and then form another loop around the ellipsoid. Normally, the geodesic line lies between the plane curves and divides the included angle in the ratio of approximately 2 to 1 as indicated by Fig. 4–4.

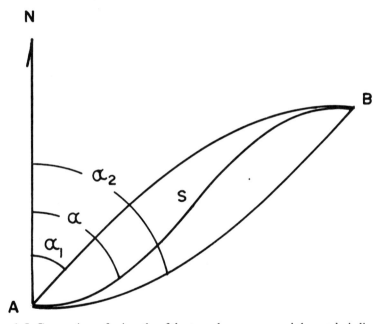

Figure 4–5. Comparison of azimuths of the two plane curves and the geodesic line.

The numerical relationships among the geodesic line and the two plane curves can be shown with reference to Fig. 4–5. The azimuth of the geodesic line AB is designated α. The azimuth of the northernmost plane curve is α_1, and that for the lower plane curve is α_2. The following relationships exist:

$$\alpha_2 - \alpha_1 = \frac{\rho}{4r^2}\ e^2 S^2 \cos^2 \theta \sin 2\alpha, \qquad (4\text{–}8)$$

$$\alpha_2 - \alpha = \frac{\rho}{12r^2}\ e^2 S^2 \cos^2 \theta \sin 2\alpha, \qquad (4\text{–}9)$$

$$\alpha - \alpha_1 = \frac{\rho}{6r^2}\ e^2 S^2 \cos^2\theta \sin 2\alpha, \qquad (4\text{-}10)$$

where ρ is the number of seconds in a radian, S is distance between A and B, and all other terms are as previously defined. All linear measurements must be expressed in consistent units. The answer is in seconds of arc.

An actual problem with $S = 1000$ km, $\theta = 45°$, and $\alpha = 45°$ gives a value of $\alpha_2 - \alpha = 1''4.$

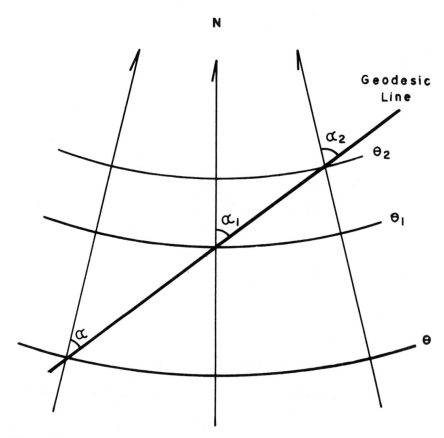

Figure 4–6. The geodesic line.

In Fig. 4–6, the geodesic line is shown cutting three meridians. Because of convergence of the meridians, as pointed out earlier in this chapter, the geodesic line intersects each successive meridian at a different angle. Let α, α_1, and α_2 be the azimuth, and θ, θ_1, and θ_2 the reduced latitude, at each of the three meridians where the geodesic line intersects. The equation of the geodesic line is

$$\cos \theta \sin \alpha = \cos \theta_1 \sin \alpha_1 = \cos \theta_2 \sin \alpha_2 = \text{a constant.} \qquad (4-11)$$

At the point where the geodesic line crosses the equator, θ is $0°$ and $\sin \alpha$ equals the constant. At the point where $\alpha = 90°$, the constant becomes the cosine of the latitude. This shows that a geodesic line which cuts the equator at any azimuth α goes northward to the parallel of latitude having a numerical value of $90°$ minus α, but will not pass beyond that limiting parallel. In the southern hemisphere, the geodesic line will likewise reach a limit of the same numerical value beyond which it cannot pass.

CALCULATION OF COORDINATES

The final products of geodesy are geodetic coordinates, distances, and directions on the reference ellipsoid. There are two general types of these computations. The first is the direct case, often referred to as Case I. The second is the inverse case, sometimes called the reverse case and often Case II. Neither case is simple. The equations are complex and difficult to derive; their derivation is beyond the scope of an introductory textbook. The computations are correspondingly long and involved, even when done with a good desk computer. The advent. of the high-speed general-purpose digital computer has removed much of the tedium. Once a computer program has been written and debugged, the solution is effortless and fast.

There are many equations for the solution of both the direct and inverse cases. Not one of these equations is really simple; they have varying degrees of complexity, depending on the length of the line and the accuracy needed. Every survey agency has its own methods and operating procedures. The equations are carefully broken down and the parts entered on a standard form so that the actual solution follows a well-defined step-by-step procedure. The authors have supervised computer offices where few, if any, of the technicians had even a vague idea of the theory behind the computations. Yet they produced consistent and good results because they followed the form.and its detailed instructions.

In approaching this chapter, the authors concluded that detailed derivations are inappropriate. Therefore, one representative set of equations will be stated and explained for each case. For the reader who wishes to dig deeper, manuals can be obtained from the government survey agencies, particularly the U.S. Coast and Geodetic Survey. Typical of these manuals are references [4–7]. Many solutions for lines from a few tens of miles in length up to super-long intercontinental lines are available in advanced textbooks [1–3].

In the direct case, we know the coordinates of one point, designated station 1, and the distance and azimuth to a new point, designated station 2, for which we wish to know the coordinates. We must also know which reference ellipsoid we shall use in our computations.

The following terms are known: a, the semimajor axis of the reference ellipsoid; f, the flattening of the reference ellipsoid; ϕ_1, the latitude of the known point; λ_1, the longitude of the known point; S, the distance between the two points; and α_{12}, the azimuth of the line measured at the known point.

The parameters a and f furnish a mathematical surface upon which to compute; ϕ_1 and λ_1 give a starting point for the computation; S provides a distance along which to compute; and α_{12} gives a direction.

The three elements to be determined are ϕ_2, the latitude of the new point; λ_2 the longitude of the new point; and α_{21}, the back azimuth of line S measured at station 2.

The first thing to remember is, Always draw a sketch! After the long and tedius computations are performed, simple mistakes, or improper

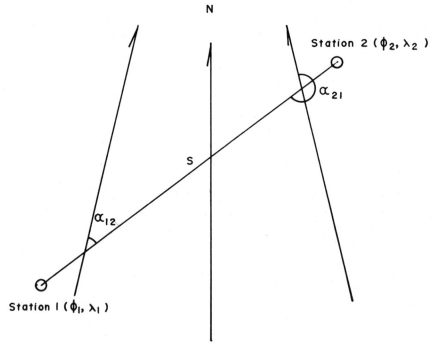

Figure 4–7. Elements of the direct and the inverse computations.

use of signs, creep in because of the lack of a sketch. A simple diagram showing the relationship between the two stations, together with a list of the known and unknown elements, will help avoid these careless errors.

Figure 4–7 illustrates the direct case. Station 1 is the point of known coordinates. The forward azimuth to station 2 is known, as well as the distance. Station 2, which is located northeast of station 1, is the station for which we need the coordinates and back azimuth.

In summary, these conditions exist:

Given	*To find*
a	ϕ_2
f	λ_2
ϕ_1	α_{21}
λ_1	
S	
α_{12}	

The set of equations selected for explanation of the direct case is that known as Puissant's Coast and Geodetic Survey formula. It is highly accurate for distances up to 60 miles, but rapidly goes wrong beyond that. For a distance of 170 miles at latitude 60°, it has an accuracy of 1 in 25,000. The equation for the difference of latitude between stations 1 and 2 is

$$\Delta\phi = SB \cos \alpha - S^2 C \sin^2 \alpha - D(\Delta\phi)^2 - hS^2 E \sin^2 \alpha, \qquad (4\text{--}12)$$

where $\Delta\phi$ is the difference in latitude in seconds of arc; S is the distance between stations 1 and 2; α is the azimuth of line 12 measured at station 1 with respect to true north;

$$h = SB \cos \alpha,$$

$$B = \frac{1}{R \sin 1''},$$

$$C = \frac{\tan \phi}{2RN \sin 1''},$$

$$D = \frac{3e^2 \sin \phi \cos \phi \sin 1''}{2(1 - e^2 \sin^2 \phi)},$$

and

$$E = \frac{(1 + 3 \tan^2 \phi)(1 - e^2 \sin^2 \phi)}{6a^2}.$$

The functions B, C, D, and E have been computed and are listed in the manuals of latitude functions referred to in Chapter 2. The latitude of station 1, the known station, is to be used as the value for obtaining these functions. It may be noted in Eq. (4–12) that, while $\Delta\phi$ is the unknown, it appears as a known in the third term on the right-hand side of the equation. For this value of $\Delta\phi$, use the sum of the first, second, and fourth terms.

The equation for the difference in longitude between stations 1 and 2 is

$$\Delta\lambda = SA \sin \alpha \sec \phi_2, \tag{4–13}$$

where $\Delta\lambda$ is the difference in longitude in seconds of arc; α and S are the same as in Eq. (4–12); ϕ_2 is the latitude of the unknown station 2 obtained from solving Eq. (4–12); and A is $1/(N \sin 1'')$.

The function A is found in the manuals of latitude functions mentioned above. Its value must be determined for the latitude of station 2 after Eq. (4–12) is solved.

The difference in azimuth is

$$\Delta\alpha = \Delta\lambda \sin \phi_m \frac{\sec \Delta\phi}{2} + (\Delta\lambda)^3 \cdot F, \tag{4–14}$$

which the reader will recognize as being identical to the value of meridian convergence expressed in Eq. (4–6). This, of course, is correct because the difference in azimuth between 1 and 2 is the amount of the convergence of the two meridians containing stations 1 and 2.

Equations (4–12), (4–13), and (4–14) are the three working equations for determining the differences in latitude, longitude, and azimuth between any two points. The coordinates of the new point and the back azimuth are simply

$$\phi_2 = \phi_1 + \Delta\phi, \tag{4–15}$$

$$\lambda_2 = \lambda_1 + \Delta\lambda, \tag{4–16}$$

$$\alpha_{21} = \alpha_{12} + \Delta\lambda + 180°. \tag{4–17}$$

In applying these last three equations, draw a sketch similar to Fig. 4–7. If the line is drawn in its approximate orientation, it will be apparent immediately whether the corrections have been applied correctly or erroneously. Furthermore, coordinates of a new point should always be computed from two known points. This is the strength of the triangulation system.

THE INVERSE CASE

The inverse case, as the name suggests, is almost the reverse of the direct case. In this case, we know the latitude and longitude of two known points and wish to find the distance and mutual azimuths between these points. With reference to Fig.4–7, the following parameters are known: a, the semimajor axis of the reference ellipsoid;

f, the flattening of the reference ellipsoid; ϕ_1, the latitude of point 1; λ_1, the longitude of point 1; ϕ_2 the latitude of point 2; and λ_2, the longitude of point 2.

This information furnishes us with a mathematical surface upon which to compute and two points of known location. We wish to find S, the distance between the two points; α_{12}, the azimuth of the line S measured at station 1 with respect to true north; and α_{21}, the azimuth of the line S measured at station 2 with respect to true north.

Although there are many solutions to the inverse problem, the one which will be explained is a representative method. It is the reverse of Puissant's Coast and Geodetic method explained for the direct case.

For convenience, let $x = S \sin \alpha$; then from Eq. (4–13) we have

$$x = S \sin \alpha = \frac{\Delta \lambda \cos \phi_2}{A}. \tag{4–18}$$

Also for convenience, let $y = S \cos \alpha$; then from Eq. (4–12) we get

$$y = S \cos \alpha = \frac{1}{B} [\Delta \phi + Cx^2 + D(\Delta \phi)^2 + E(\Delta \phi) x^2]. \tag{4–19}$$

Dividing Eq. (4–18) by Eq. (4–19) gives

$$\frac{x}{y} = \frac{S \sin \alpha}{S \cos \alpha} = \tan \alpha, \tag{4–20}$$

after which α is found. This is actually the azimuth measured at point 1, or α_{12}. To obtain α_{21}, the value of $\Delta \alpha$ must be computed by Eq. (4–14) and applied as in Eq. (4–17).

The value of S is given by either

$$S = x \operatorname{cosec} \alpha, \tag{4–21a}$$

or

$$S = y \sec \alpha. \tag{4–21b}$$

The value of A in the inverse solution is taken at the latitude of station 2, whereas the values of B, C, D, and E are for the latitude of station 1. These values are all listed in the manuals of latitude functions mentioned previously.

For both the direct case and the inverse case, the unit of measurement for $\Delta \phi$, $\Delta \lambda$, and $\Delta \alpha$ is seconds of arc. The designation has been omitted for simplicity. Once the values in seconds have been computed, they are converted to the standard notation of degrees, minutes, and seconds before they are applied as corrections. The value of S may be in any linear unit of measurement. Normally, the meter is used. With one exception, the manuals of latitude functions use the meter as the unit of measurement. The exception is the Everest ellipsoid, which uses the value of the Indian yard. In converting to feet, statute miles, nautical miles, or some other unit of measurement, it is important that the correct conversion factor be used. While the internationally adopted length of the meter is constant, and hopefully will remain that way, the other values may change. The currently accepted values, adopted by the United States Bureau of Standards in international agreement, became effective on 1 July 1959. The values most frequently used in geodesy are listed in Appendix B.

REFERENCES

1. G. Bomford, *Geodesy*. The Clarendon Press, Oxford, 1952.
2. D. Clark and J. Clendinning, *Plane and Geodetic Surveying, Vol. Two, Higher Surveying*, 4th Ed. Constable, London, 1951.
3. G. L. Hosmer, *Geodesy*, 2nd Ed. Wiley, New York, 1930.
4. U.S. Coast and Geodetic Survey, Formulas and Tables for the Computation of Geodetic Positions, *Spec. Publ. No.* **8**, 7th Ed. Government Printing Office, Washington, D.C.
5. U.S. Coast and Geodetic Survey, Formulas and Tables for the Computation of Geodetic Positions on the International Ellipsoid, *Spec Publ. No.* **200**, Government Printing Office, Washington, D.C.
6. U. S. Coast and Geodetic Survey, Manual of Triangulation Computation and Adjustment, *Spec. Publ. No.* **138**, Government Printing Office, Washington, D.C.
7. U.S. Coast and Geodetic Survey, Manual of Plane-Coordinate Computation, *Spec. Publ. No.* **193**, Government Printing Office, Washington, D.C.

Chapter 5   Geodetic Astronomy

Geodetic astronomy is the application of astronomy to the precise determination of latitude, longitude, and azimuth on the earth.

A distinction must be made between the two terms "geodetic positions and directions" and "astronomical positions and directions." All the discussion about latitude, longitude, and azimuth encountered so far in this book has been geodetic in nature because the three quantities have been referred to the ellipsoid, which is the mathematical figure of the earth. Astronomical positions and directions are determined with reference to the geoid. The difference between the two is a function of the deflection of the vertical which has been described very briefly earlier and is discussed in detail in Chapter 8. The manner in which the geodetic and the astronomical quantities are related will be shown later in this chapter. In order to avoid confusion, the geodetic latitude, longitude, and azimuth are shown as ϕ, λ, and α, respectively, which has been the practice heretofore in this book. The astronomical latitude, longitude, and azimuth bear the corresponding capital Greek letters Φ, Λ, and A.

The determination of astronomical positions and directions is essentially a study in applied spherical trigonometry. The basic equations of spherical trigonometry are shown in Appendix A. In contrast with geodesy, which has complicated spheroidal triangles, astronomy deals with true spherical triangles. The celestial sphere is considered to be a vast sphere of infinite dimensions with the observer at the center.

Before detailed derivations for latitude, longitude, and azimuth are attempted, it is necessary to have an understanding of some of the fundamental properties associated with geodetic astronomy. The definitions which follow will have more meaning as they are related to appropriate figures and derivations.

Of the four astronomical systems of coordinates in common use, two are important to the geodesist. These two are the horizon system and the equator system. They differ in the position of the fundamental circle, an arbitrary great circle of the celestial sphere.

The horizon system is shown in Fig. 5–1. The fundamental circle is the horizon. Small circles parallel to the horizon are called almucan-

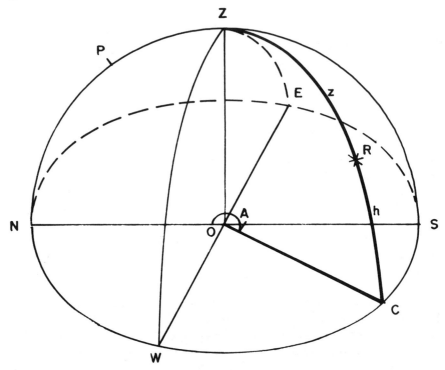

Figure 5–1. The horizon system of coordinates.

tars. The zenith, designated Z, is the point at which a plumbline produced upward from the observer's position O intersects the celestial sphere. The diametrically opposite point on the celestial sphere is the nadir. The zenith and the nadir are the poles in the horizon system. Great circles which pass through the zenith and the nadir are called vertical circles; they are perpendicular to the horizon. The vertical circle through the star R is ZRC. The astronomical azimuth A of the star is the angular distance measured clockwise on the horizon from the north point to point C which is the foot of the vertical circle through R. The altitude h of the star is the angular distance measured upward from the horizon on the vertical circle through the body. It is the arc CR. The zenith distance z of the star R is the arc ZR; it is the complement of the altitude, or $z = 90° - h$. The two quantities azimuth and altitude (or zenith distance) determine the position of a heavenly body in the horizon system at any given instant.

The horizon system is used for some geodetic purposes; however, it has its weak points. The coordinates of a heavenly body are not constant. The altitude and azimuth of a star continuously change because of diurnal motion. Furthermore, the system is local. The altitude and

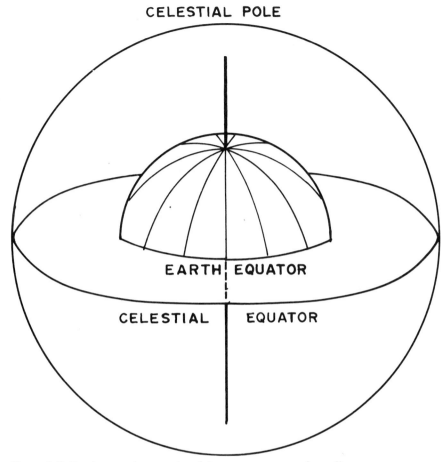

Figure 5–2. Fundamental properties of the equator system of coordinates.

azimuth of a star are not the same for two observers at different locations.

The second astronomical coordinate system of geodetic significance is the equator system. It is shown in Fig. 5–2. The celestial poles, hereafter referred to simply as the poles, are the two points where the axis of rotation of the earth, if extended, would intersect the celestial sphere. The celestial equator is the intersection of the celestial sphere and the plane of the earth's equator. It is the fundamental circle of the equator system. Small circles parallel to the celestial equator are known as parallels of declination and are equivalent to parallels of latitude on the earth.

The vertical circle which passes through the celestial poles is known as the celestial meridian or, more simply, as the meridian. The prime vertical is the vertical circle at right angles to the meridian. The great circles which are perpendicular to the equator are known as hour circles. The hour circle through the zenith is the meridian of the observer. It is shown by PZP' in Fig. 5–3. Thus the meridian of the observer, which is a vertical circle in the horizon system, is also an hour circle in the equator system. The meridian zenith distance, z_m, of a heavenly body is the angular distance from the zenith to the body, measured along the hour circle through the body.

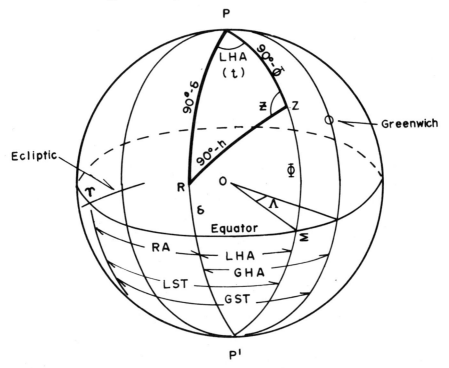

Figure 5–3. The equator system of coordinates.

Figure 5–3 also shows additional properties of the equator system. The local hour angle (LHA) of a heavenly body R is the angular distance measured westward on the equator from the point of intersection of the equator and the meridian of the observer (the Σ point), to the foot of the hour circle through the body. What is exactly the same thing, it is the angle at the pole from the observer's meridian westward to the hour circle through the body. The meridian angle t of a heavenly body is the angular distance measured either eastward or westward, whichever is the shorter distance, on the equator from the Σ

point to the foot of the hour circle through the body. LHA and t are similar. The former is measured only in the westward direction; the latter is measured either east or west from 0° to 180° and carries the suffix E or W. In the case of star R in Fig. 5–3, LHA and t have the same value. If R had been to the east of the observer's meridian, t would have been measured in that direction, whereas the LHA would have been measured westward from the meridian of the observer all the way around the equator to the foot of the hour circle through R.

The Greenwich hour angle (GHA) of a heavenly body is the angular distance measured westward on the equator from the meridian of Greenwich to the foot of the hour circle through the body.

A point of great importance in geodetic astronomy is the vernal equinox, designated ♈ (the Ram's head), and often called the first point of Aries. It is that intersection of the ecliptic with the equator at which the sun crosses the equator from south to north. This occurs only once each year, on or near 21 March, hence the name vernal.

The right ascension (RA) of a body is the angular distance measured eastward on the equator from the vernal equinox to the foot of the hour circle through the body. Right ascension is usually expressed in units of time: hours, minutes, and seconds.

Local sidereal time is the local hour angle of the vernal equinox. By reference to Fig. 5–3, it can be seen that LST = LHA + RA. Greenwich sidereal time is the Greenwich hour angle of the vernal equinox, or GST = GHA + RA.

The declineation δ is the angular distance from the equator to a heavenly body measured on the hour circle through the body. It carries a positive sign when the body is north of the equator, and a negative sign when it is south of the equator.

The azimuth angle of a heavenly body is the angular distance measured at the observer's position eastward or westward on the horizon to the foot of the vertical circle through the body. It is designated by the symbol $\mathcal{Z}$ and should not be confused with $\mathcal{Z}$, which is the zenith, or with z, which is zenith distance.

In the equator system, the coordinates of a heavenly body are expressed either by hour angle and declination, or by right ascension and declination. Because right ascension and declination change very slowly through the years. they are especially useful for identifying the position of a body on the celestial sphere.

Knowledge of the fundamental properties of astronomy, as defined above and as illustrated by Figs. 5–1, 5–2, and 5–3, is absolutely essential to a proper understanding of the means by which latitude, longitude, and azimuth are determined. The end products of geodetic astronomy are latitude, longitude, and azimuth. Some instruments are

peculiarly adapted to the determination of one quantity, such as the
transit instrument for longitude and the zenith telescope for latitude;
some are adapted to the simultaneous determination of both latitude
and longitude, such as the zenith camera and the astrolabe; others,
particularly the theodolite, can be used for the determination of all
three quantities.

THE DETERMINATION OF LATITUDE

The astronomical latitude Φ of an observer is the angle between the
direction of a plumbline at that point and the plane of the earth's
equator. It may be either north or south depending on the position of
the observer with respect to the equator. Another definition is that
latitude is the declination of the zenith.

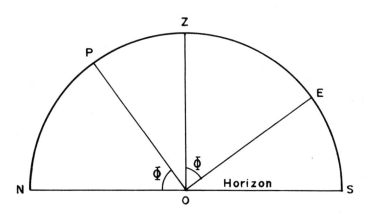

Figure 5–4. Astronomical latitude.

Figure 5–4 is taken in the plane of the meridian. The observer is at
point O. The north and south points on the horizon are N and S,
respectively. P, the celestial pole, is at right angles to the equator,
designated E. The zenith of the observer is Z. By definition, the latitude
Φ is the declination of the zenith, or the angular distance from the
equator to the zenith measured along the hour circle. Since ZO is
perpendicular to the horizon NS, and OP is perpendicular to the
equator OE, it follows that $\angle NOP$ is equal to Φ. From this basic geometry
comes the fundamental fact that the altitude of the pole is equal to the
latitude of the observer. This fact is often used in simple determinations
of latitude from sightings on Polaris followed by a correction for the

local hour angle of observation, which is found in the American Ephemeris and Nautical Almanac [4], hereafter referred to simply as the Almanac. A correction for atmospheric refraction is necessary also. The result is not of geodetic quality.

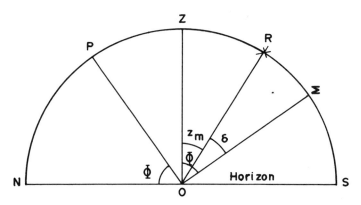

Figure 5–5. Relationships among latitude, meridian zenith distance, and declination for a body south of the zenith.

Another method for determining latitude is to measure the meridian altitude or its complement, the meridian zenith distance, of a star. Figure 5–5 is drawn in the plane of the meridian of the observer. R is a star at the instant of crossing the meridian. The Σ point is the intersection of the observer's meridian and the equator. From this sketch, we see that $\Sigma Z = \Phi$, the astronomical latitude of the observer; $R Z = z_m$, the meridian zenith distance of the star; and $\Sigma R = \delta$, the declination of the star.

From these relationships, we can write

$$\Phi = z_m + \delta \quad \text{(for stars south of the zenith).} \tag{5–1}$$

When a star is chosen which has a declination greater than the observer's latitude, it will cross the meridian north of the zenith. Figure 5–6 shows the conditions when the star is above the pole and north of the zenith. From the figure, we may write

$$\Phi = \delta - z_m \quad \text{(for upper transits north of the zenith).} \tag{5–2}$$

When the star crosses the meridian of the observer below the pole, as

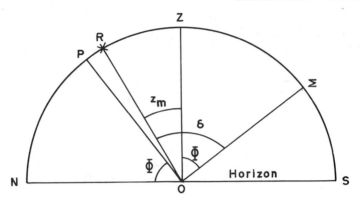

Figure 5–6. Relationships among latitude, meridian zenith distance, and declination for a body north of the zenith and above the pole.

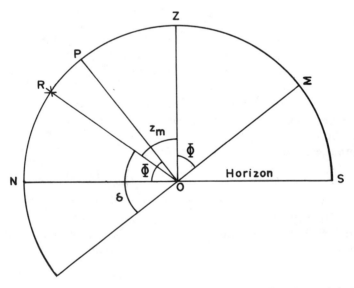

Figure 5–7. Relationships among latitude, meridian zenith distance, and declination for a body north of the zenith and below the pole.

in the case of circumpolar stars at lower transit, Fig. 5–7 shows the geometry. In this case, the latitude is given by

$$\Phi = 180° - z_{\mathrm{m}} - \delta \quad \text{(for stars at lower transit).} \tag{5–3}$$

These three equations (5–1), (5–2), and (5–3) also hold true when the observer is in the southern hemisphere except that in Eq. (5–1) the word south should be replaced by north, and in Eq. (5–2) the word north should be repaced by south. Equation (5–3) holds true as written. These relationships may be proved by drawing simple sketches of the southern hemisphere.

Care must be used in choosing the sign of δ when these equations are used. When the declination is in the latitude of the observer, its sign is positive; when the declination is in the latitude opposite the observer, its sign is negative.

In applying this method, the selected star is tracked for several minutes before transit, which is defined as its passage across the meridian of the observer. The tangent screw of the vertical circle is used to keep the horizontal wire on the star. When the star appears to move along the horizontal wire with no apparent vertical motion, its meridian zenith distance is recorded. If the meridian altitude is read instead of the meridian zenith distance, it is subtracted from 90° to obtain the meridian zenith distance z_{m}. The declination of the star being observed is found in the Almanac. Finally, one of Eqs. (5–1), (5–2), or (5–3) is used, depending on the conditions of observation. Of course a correction for atmospheric refraction must be applied.

Several other methods are available for determining latitude, such as observing the altitude of a star not on the meridian when the time is known, and using circummeridian altitudes. From any of these methods it is possible to determine the precise latitude of a place provided we know the proper atmospheric refraction correction to apply. Herein lies the weakeness of these methods discussed previously. Because of atmospheric refraction, a ray of light from a celestial body undergoes a change in direction as it penetrates the atmosphere. When the ray of light from the heavenly body R in Fig. 5–8 arrives at the observer's position, it will appear to come from R', which is the final direction of the light as it enters the eye of the observer. Therefore, any object will appear at a higher altitude than it actually is. The amount of this angular displacement in altitude is the refraction correction r which must be subtracted from the observed altitude, or added to the observed zenith distance, in this manner:

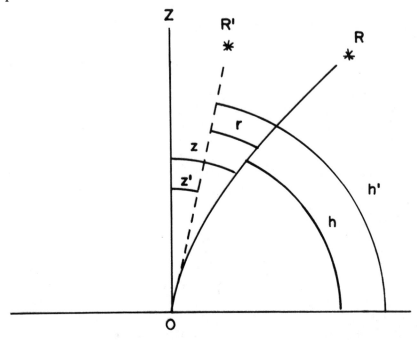

Figure 5–8. Refraction of light in the earth's atmosphere.

$$h = h' - r \tag{5-4a}$$

or

$$z = z' + r, \tag{5-4b}$$

where h' is the observed altitude and z' is the observed zenith distance.

The magnitude of r may be obtained from tables and formulas. One such correction is

$$r = 60.6 \cot h', \tag{5-5}$$

where r is the refraction correction in seconds of arc.

The error in this correction increases as the altitude decreases and may reach 10 seconds of arc at low altitudes. For this reason, a more accurate correction is needed. Because the index of refraction varies with the temperature and pressure of the atmosphere, it is necessary to use an equation written in terms of these quantities. One equation, known as Comstock's empirical formula, is

$$r = \frac{983b}{460 + t} \cot h' \tag{5-6}$$

where r is the refraction correction in seconds of arc; b is the barometric pressure in inches of mercury; and t is the temperature in degrees Fahrenheit.

For altitudes greater than $15°$ above the horizon, the error produced by this equation will seldom exceed 1 second of arc. For some types of surveying, this error is not unreasonable. In geodetic surveying, how-ever, it is intolerable. One second of arc in latitude corresponds to slightly more than 100 feet on the earth's surface. Therefore, for geodetic purposes, we must seek a better method of determining lati-tude than those previously discussed.

This better method is the Horrebow-Talcott method. In this pro-cedure, we observe a series of stars to the north and to the south of the zenith. Each pair of stars consists of a north star and a south star of known declination, at approximately equal zenith distances and transiting the meridian at about the same time. Since Eqs. (5–1) and (5–2) were written for stars to the south of the zenith and stars to the north of the zenith, respectively, their combination is used in the Horrebow-Talcott method.

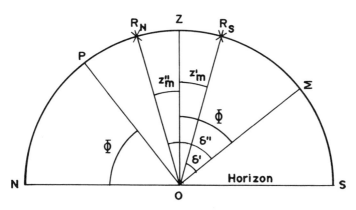

Figure 5–9. Determination of latitude by the Horrebow-Talcott method.

Figure 5–9 shows the plane of the observer's meridian. R_s is the star transiting the meridian to the south of the zenith. Its declination is δ' and its meridian zenith distance z'_m. R_n is the star which transits the meridian to the north of the zenith. Its declination is δ'' and its meridian zenith distance z''_m. The superscripts are prime and double-prime and not minutes and seconds of arc. Equations (5–1) and (5–2) may be rewritten for this special case as follows:

$$\Phi = \delta' + z'_m \quad \text{and} \quad \Phi = \delta'' - z''_m.$$

Adding these two equations gives

$$2\,\Phi = \delta' + \delta'' + z'_m - z''_m \tag{5-7}$$

or, expressed differently,

$$\Phi = \tfrac{1}{2}(\delta' + \delta'') + \tfrac{1}{2}(z'_m - z''_m). \tag{5-8}$$

This equation indicates that the only observed quantity needed is the small difference in meridian zenith distances between the pair of stars and not the absolute meridian zenith distances of the two stars. The declinations of the two stars, needed to complete the information for Eq. (5–8), are found in the Almanac.

This method may be used with several different instruments, among which are the observatory-mounted zenith telescope, the Bamberg

Figure 5–10. Bamberg broken-telescope transit (U.S. Coast and Geodetic Survey).

broken-telescope transit, shown in Fig. 5–10, and the Wild T–4 Universal Instrument, which is also a broken-telescope instrument, shown in Fig. 5–11. The term "broken telescope" is used to describe a telescope having a prism in the optical axis which directs the rays of light at right angles through the hollow axis of the instrument to the eyepiece. With this arrangement, the observer remains in the same position no matter what the zenith distance of the observed body may be.

Figure 5–11. Wild T–4 Universal Instrument (Wild Heerbrugg Instruments, Inc.).

These instruments are equipped with an eyepiece micrometer having a movable reticle. This mechanism is designed to measure to a fraction of a second of arc the difference in meridian zenith distances of a pair of stars. The field of view depends on the instrument used; in the case of the Wild T–4, it is 30 minutes of arc. Because only the difference of two meridian zenith distances is read instead of the absolute quantities, this small field of view is adequate.

The horizontal axis of the telescope is oriented in a true east-west direction so that the line of sight coincides with the meridian when the telescope is rotated about that axis. Reversing the telescope 180° about the vertical axis again orients the instrument so that rotation about the horizontal axis describes the meridian on the other side of the zenith. Without resetting the telescope, and merely by reversing the instrument about its vertical axis, one may observe another star of approximately the same meridian zenith distance on the opposite side of the zenith.

Prior to the field observations, an observing list must be compiled. The limited number of stars shown in the Almanac is inadequate. A general star catalogue is needed. One catalogue is the *Apparent Places of Fundamental Stars* [3], which contains the positions of 1535 stars, with 26 of these being north circumpolar stars and 26 south circumpolar stars. Another listing is the *Boss Star Catalogue* [1], which provides the mean positions of 33,342 stars for a given date. The mean position of a star for the given date must be reduced to its apparent place at any given instant by use of equations provided in the Almanac. It is important to remember that in geodetic work the apparent place of a star is always used.

It would, of course, be a long and tedious process to search through a star catalogue to find six to eight pairs of stars which are constrained to meeting certain definite conditions. The difference in time between the transit of each star must be at least 4 minutes to allow for reading the first star, reversing the instrument, and reading the second star. The time should not exceed 20 minutes because the constants of the instrument might change if the observing period is unduly long. Stars having meridian zenith distances greater than 30° should not be used. The extent of the field of the micrometer is generally about 30 minutes of arc and so the difference in the meridian zenith distance of a pair cannot exceed that amount. Stars fainter than eighth magnitude should be disregarded.

With these constraints, the selection of an observing list must be made methodically. Since we know the time period of our observation, we must choose stars having a right ascension corresponding to this time. The declination must, as stated above, be chosen so that the meridian zenith distance does not exceed 30°. Since we have some prior knowledge of the latitude of the observer, we can use this best estimate and rearrange Eq. (5–7) into the form

$$(\delta' + \delta'') - 2\Phi = z_m'' - z_m'. \tag{5–9}$$

This equation says that the sum of the declinations of the two stars minus twice the latitude is equal to the difference in the meridian zenith distances. This difference is limited by the field of view of the telescope and the extent of the micrometer, which in the case of the Wild T–4 is 30 minutes of arc.

Entering the catalogue with the selected period of right ascension, we select one star brighter than eighth magnitude which has a declination so that the meridian zenith distance does not exceed 30°. Then searching in the catalogue not less than 4 minutes nor more than 20 minutes of right ascension later, we choose a second star of sufficient brightness which fulfills the conditions of Eq. (5–9). A little practice will enable the observer to compile a list of six or eight pairs of stars without difficulty.

Prior to observation of the first star in a pair, the telescope is set to the average meridian zenith distance of the two stars. A few minutes before the time of transit, the first star will come into the field of view of the telescope and appear to travel parallel to the movable wire in the micrometer. The micrometer head must then be turned until the movable wire bisects the star as it transits the meridian. The micrometer reading is recorded; also, the time of transit is recorded merely as a check to insure that the proper star was observed. Without disturbing the setting of the telescope, merely reverse it about the vertical axis until it hits the 180° stop. Proceed with the second star in an identical manner. Knowing the value of one turn of the screw for the specific instrument enables the difference in micrometer readings between the first and the second observations to be translated directly into seconds of arc. This information completes the needs for solution of the astronomical latitude by use of Eq. (5–8).

Refraction, the nemesis of all astronomical observations, is eliminated effectively by this method. For one thing, stars having meridian zenith distances greater than 30° are not used. More importantly, only the small difference in meridian zenith distances of a pair of stars at nearly equal distances north and south of the zenith enters into the computations. Therefore, the refraction effect is very nearly canceled. Accuracies of 0.1 second of arc can be obtained.

Of the several methods of latitude determination described above, only the Horrebow-Talcott method employing the zenith telescope or similar instrument furnishes the precision needed in geodetic work. The other methods have been included to provide a background for the reader and to suggest procedures which may be used when a less precise preliminary determination is needed.

THE DETERMINATION OF LONGITUDE

Geodetic longitude is the angular distance measured on the terrestrial equator from the intersection of a fixed meridian and the equator to the foot of the meridian through the observer's position. The meridian through Greenwich, England, is taken as the fixed meridian. Longitude is reckoned positive westward and negative eastward from the meridian of Greenwich. The standard designation of geodetic longitude is λ.

Astronomical longitude differs from geodetic longitude by the longitudinal component of the deflection of the vertical. The two are related in a manner which will be explained at the end of this chapter. Astronomical longitude is designated Λ.

Longitude is correlated directly with time. The difference in longitude between any two places is equal to the difference in their local times. This local time may be sidereal, apparent, or civil; in geodetic observations, sidereal time is normally used.

In Fig. 5–12, G designates Greenwich. The point M, the observer's position, is west of Greenwich and is the point for which the longitude is desired. The angle Λ is the difference in astronomical longitude between the Greenwich meridian and the meridian through M. With the vernal equinox at ♈, the sidereal clock at Greenwich reads Greenwich sidereal time (GST); the sidereal clock at M reads local sidereal time (LST). The angle Λ is the difference between the two, or

$$\Lambda = \text{GST} - \text{LST}. \qquad (5\text{–}10)$$

This equation expresses Λ in terms of time. The following relationships are used to convert units of time to units of arc:

$$
\begin{aligned}
1 \text{ hour of time} &= 15° \text{ of arc,} \\
1 \text{ minute of time} &= 15' \text{ of arc,} \\
1 \text{ second of time} &= 15'' \text{ of arc.}
\end{aligned}
$$

In theory, the determination of longitude is quite simple. In practice, it is a difficult and painstaking task, particularly in the precise measurement of time. The determination of the longitude of a place is a three-step solution:

1. *Determine Greenwich sidereal time at the instant of transit of a star.* This is done by use of a sidereal chronometer which is set to Greenwich sidereal time. At the instant of transit, a chronographic record is made.

2. *Determine the local sidereal time at the point M where the longitude is desired.*

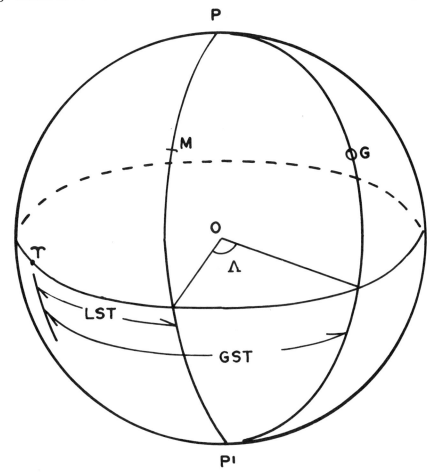

Figure 5–12. The relationship between time and longitude.

This is done by obtaining the right ascension for the star which was observed in transit in step 1. Since local sidereal time is defined as the right ascension of the observer's meridian, it follows that, when the star is on the meridian, the local sidereal time is equal to the right ascension of the star. This can be seen from the basic equation of sidereal time,

$$LST = RA + LHA, \qquad (5\text{--}11)$$

where LST = local sidereal time, RA = right ascension, and LHA = local hour angle. In the case of a star on the meridian, the local hour angle is zero. Therefore,

$$LST = RA. \qquad (5\text{--}12)$$

The RA of a star, and hence its LST, is taken from the Almanac or a star catalogue.

3. *Compare the Greenwich sidereal time as determined by step* 1 *with the local sidereal time as explained in step* 2. The difference is the longitude of the place, as shown by a return to Eq. (5–10).

The instrument most generally used in the precise determination of longitude is the transit instrument mounted in an observatory or on a concrete pedestal. This precise instrument should not be confused with the ordinary surveyor's transit used in plane surveying. It consists basically of a telescope, a horizontal axis perpendicular to the telescope, and a reticle in the focal plane. The horizontal axis is the axis of rotation and is mounted in an east-west direction which causes the telescope to move in the plane of the meridian. With the instrument in perfect adjustment, the mean wire of the reticle traces the meridian when the telescope is rotated about its horizontal axis. A graduated circle is attached so that the proper meridian zenith distance may be set to assist in locating the stars in an observing list.

Other instruments used extensively in the determination of longitude are the Bamberg broken-telescope transit shown in Fig. 5–10 and the Wild T–4 Universal Instrument shown in Fig. 5–11. These instruments are similar to the transit instrument but are more convenient to use. The horizontal axis is hollow. A 90° prism in the optical axis of the telescope directs the rays through the hollow axis to the eye. All observations may be made from one position.

A time signal from the U.S. Bureau of Standards through station WWV is recorded before the observations, again after 6 stars have been observed, and a third time after the operation is completed. From these time signals, the rate of the chronometer is determined. The chronometer drives a cylinder so that the beat of the clock is transmitted by an electric circuit to an electromagnet which releases a pen and marks the seconds on the sheet of graph paper which is wrapped around the cylinder. Other chronographs use a reel and paper tape in lieu of the cylinder and graph paper. As a star moves across the field of the telescope, it is continuously bisected with a movable wire by turning the handwheels of an impersonal micrometer. As the wheels are turned, electrical contacts are broken at regular intervals. These contacts are recorded on the chronograph. The readings of the chronograph, which can be estimated to 0.01 seconds of time, are averaged. The time thus obtained is the instant of transit of the star over the fictitious mean wire. Figure 5–13 shows a combination time-signal receiver, chronograph, and chronometer.

The observed time must be corrected for diurnal aberration. Since

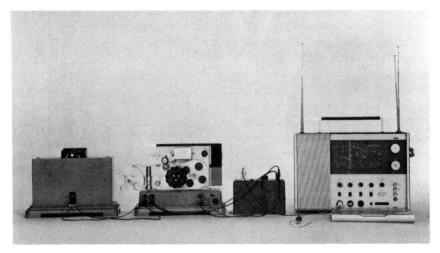

Figure 5–13. From left to right: Chronometer, chronograph, and time-signal receiver (Wild Heerbrugg Instruments, Inc.).

the velocity of light is finite compared with the rotational velocity of the observer on the earth's surface, a star is apparently displaced. This displacement, known as diurnal aberration, is designated K. Its magnitude is

$$K = 0.021 \sec \delta \cos \phi, \qquad (5\text{–}13)$$

where K is the diurnal aberration in seconds of time; δ is the declination of the star observed; and ϕ is the latitude of the observer. This correction must be subtracted from the observed time of the meridian transit of the star.

A single determination of longitude consists of the transit observations of 12 stars of which 6 are south and 6 north of the zenith. One in each set of 6 should be within 8° of the zenith. The stars should be chosen and arranged in an observing list so there are at least several minutes of time between successive stars. As the star appears in the field, it is bisected by the vertical wire of the impersonal micrometer as described previously. This procedure is continued until the entire list of 6 stars is finished. The telescope is then reversed 180° about its vertical axis and the second set of 6 stars is observed.

As in an ordinary surveyor's transit, it is necessary to keep the instrument in as nearly perfect adjustment as possible. The observing list and the observing procedures are designed to minimize the errors, such as selecting the same number of stars north and south of the zenith and reversing the instrument in order to eliminate the azimuth error. Since these errors in the transit instrument can be minimized, the main element contributing to error in longitude determination is that of time. The velocity of rotation of a point on the equator is approximately 1500 feet per second. At latitude 45°, it is over 1000 feet per second. When time is interpolated to 0.01 second from the chronograph, this means a possible error of about 10 to 15 feet in longitude. The experience of the authors is that an overall accuracy of 50 to 100 feet in longitude is more realistic when the transit instrument is used with the procedures which have been outlined.

THE SIMULTANEOUS DETERMINATION OF LATITUDE AND LONGITUDE

So far in this chapter we have considered the independent determination of latitude and longitude by use of instruments designed for those specific purposes. Several instruments are available for the simultaneous determination of latitude and longitude. One of these is the astrolabe; another is the zenith camera. The astrolabe will be considered in this chapter.

For years it has been common practice at sea and in the air for navigators to use a sextant or an octant in determining the latitude and longitude of a place by observing the altitudes and corresponding times of two or more stars. Accuracies in the order of 2 minutes of arc are usually achieved. Another method, similar to that used by navigators, is often used in geodetic surveys. It is based on determining the time when a star attains a fixed altitude. Accuracies of 1 to 2 seconds of arc are obtainable. The astrolabe is the instrument normally used. It is a rather simple device for observation on the almucantar of 30° zenith distance. There are two types of astrolabes: the prismatic astrolabe and the pendulum astrolabe.

The prismatic astrolabe may be an independent unit specifically designed as such, or it may be an attachment for use on a theodolite. An attachment for use on the Wild T–3 Precision Theodolite is shown in Fig. 5–14. The prismatic astrolabe has three principal parts:

(1)   the horizontal telescope, at the focus of which is a reticle;

(2)   the 60° prism, which is located in front of the objective with one face perpendicular to the optical axis of the telescope;

(3)   a mercury trough located just below the lower face of the prism.

Figure 5–15 shows the arrangement of the components of a typical astrolabe. Two parallel rays S and S' from a star of zenith distance 30° are shown. The ray S' enters the prism at right angles to the surface AB, is reflected at surface AC, and emerges perpendicular to surface BC. The ray S is reflected by the mercury surface, enters the prism at right angles to surface AC, and then emerges perpendicular to surface BC. Therefore, the two rays emerging from surface BC are parallel and produce a single

Figure 5–14. Prismatic astrolabe attachment for use on the Wild T–3 Precision Theodolite (Wild Heerbrugg Instruments, Inc.).

image of the star at *F when the zenith distance of the star is* 30°. At zenith distances other than 30°, the observer will see two images.

The eyepiece *H* is used for finding the star; it has a field of 1°5 and a magnification of 30 diameters. The eyepiece *G* is used for normal observations and has a field of 0°5 and a magnification of 80 diameters.

The view through the eyepiece *G* is shown in sequence in Fig. 5–16. First is the view before the star attains 30° zenith distance; second, the instant the star is at 30° zenith distance; and third, after the star passes 30° zenith distance. It is better practice to set the astrolabe so that the stars do not come to coincidence but appear to pass each other a small

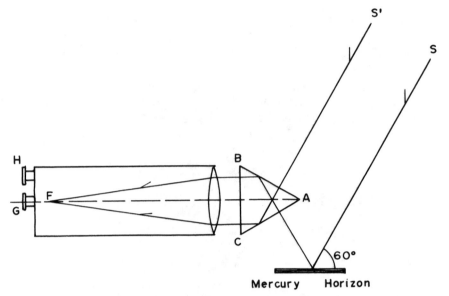

Figure 5–15. Diagram showing the optical system of the prismatic astrolabe.

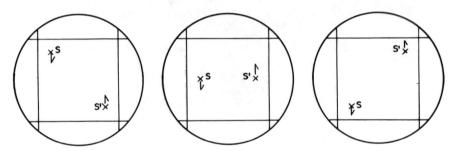

Figure 5–16. View through the eyepiece of an astrolabe showing three positions of the double image of a star.

distance apart. This can be done by rotating the telescope about its optical axis.

The pendulum astrolabe does not employ the double-image principle used in the prismatic astrolabe. With the pendulum astrolabe, or "bent telescope," only one image of a star is observed as it crosses a fixed wire of the reticle. This fixed wire represents a constant zenith distance of 30°. The nonpendulus portion is brought to within 1 minute of arc of being perfectly level. Then four thumbscrews are loosened, allowing a metallic mirror, which is attached to a pendulum, to be suspended by sensitive springs. The pendulum comes to rest within $\pm$ 0".1 from the direction of local gravity. The mirror is then perpendicular to the pendulum and therefore rests in the local horizontal plane. Rays of light enter the objective, are reflected from the mirror, and then observed through the eyepiece.

In theory, determinations of the position of an observer are identical regardless of the type of astrolabe. The theory is quite simple. From any point on the surface of the earth, the zenith distance of a star expresses the angular distance on the surface of the earth from the substellar point of the star to the observer. If the zenith distance of a star is determined, the observer is on the circumference of a circle with its center at the substellar point of the star and angular radius equal to the determined zenith distance. Observations of two stars will yield two circles which intersect. One point of intersection is the location of the observer. An observation of a third star will resolve the ambiguity.

In practice, this method of drawing circles would not be accurate

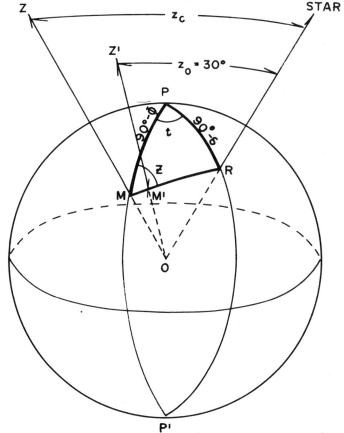

Figure 5–17. Surface of the earth showing the substellar point of the star at R and the assumed position of the observer at M.

enough for geodetic purposes. Therefore, either a mathematical solution or a graphical solution employing lines of position is used. In the authors' experience the graphical solution is less tedious and gives comparable results to that obtained mathematically. Furthermore, the graphical solution enables the reader to follow better the successive steps in the theory. For these reasons, the graphical solution will be explained.

Figure 5–17 shows the geometric conditions on the surface of the earth. Point M is the position of the observer. His position is approximately known and these assumed coordinates are known as Φ and Λ. The right ascension and declination of the star are taken from the Almanac or star catalogue for the instant of observation. The substellar point is designated R. In the spherical triangle PMR, we know the following:

$$PM = 90° - \Phi,$$
$$PR = 90° - \delta,$$
$$t = \text{the meridian angle of the star.}$$

The local chronograph is set to LST by subtracting the assumed longitude Λ from GST. Then LHA = LST − RA, from which t is found.

In the same spherical triangle PMR, we wish to find

$$MR = z_c,$$

which is the computed zenith distance, and

$$\angle PMR = \mathcal{Z}$$

the azimuth angle, which is reckoned either east or west of true north as the star is east or west of the meridian.

From the law of cosines of a spherical triangle, we may write

$$\cos z_c = \cos(90° - \Phi) \cos(90° - \delta) + \sin(90° - \Phi) \sin(90° - \delta) \cos t$$

or, simplified,

$$\cos z_c = \sin \Phi \sin \delta + \cos \Phi \cos \delta \cos t. \tag{5–14}$$

From the law of sines of a spherical triangle, we obtain

$$\frac{\sin \mathcal{Z}}{\sin(90° - \delta)} = \frac{\sin t}{\sin z_c},$$

or

$$\sin \mathcal{Z} = \frac{\sin t \cos \delta}{\sin z_c}. \tag{5–15}$$

The computation of Eqs. (5–14) and (5–15) furnishes the computed zenith distance z_c and its corresponding azimuth angle $\mathcal{Z}$. The zenith distance as actually observed is called z_0 and is designed to be 30° in the case of astrolabes. The observer is situated on the circumference of a circle with its center at the substellar point R and radius z_0. A small arc of this circle is shown intersecting MR at M'. This arc would have intersected M if the assumed position had been the actual position of the observer. This distance MM' is equal to z_c minus z_0 and is known as the intercept.

Next, this intercept is plotted on a graph to a convenient scale. The y axis of the graph represents the meridian through point M. Since the intercept is defined as the computed zenith distance z_c minus the observed zenith distance z_0, the intercept is plotted toward the substellar point when the intercept is positive; when the intercept is negative, it is plotted away from the substellar point. At the end of the intercept, a

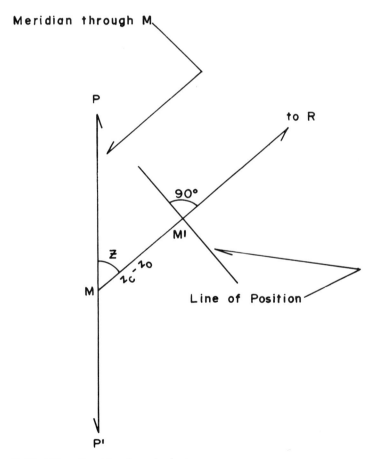

Figure 5–18. A line of position for a single star.

short straight line is drawn perpendicular to the intercept. This line represents the line of position; it is a short segment of the arc through M'. The direction to the substellar point is provided by the computed azimuth angle $\mathcal{Z}$. Figure 5–18 indicates the line of position corresponding to the conditions of Fig. 5–17. This illustrates the procedure for one star.

An observing list consists of 20 stars, 5 in each quadrant. The stars must be selected so they will intersect the circle of constant 30° zenith distance within the period of time selected for the observations. Sometimes the same star may be used twice, once when it intersects the circle of constant zenith distance east of the meridian and again at the intersection west of the meridian. For any latitude Φ of the observer, the stars which will attain a zenith distance of 30° must have a declination within the limits of Φ plus 30° and Φ minus 30°. The selection of 20 stars which fulfill these conditions and the computation of the azimuth angle and local sidereal time is a laborious process when done without some short cut. This assistance has been provided in the form of tables [2] prepared for each 1° of latitude from 60°S to 60°N. These tables give the LST and azimuth at which stars attain the 30° zenith distance. An interpolation for the assumed latitude of the observer is sufficiently accurate to provide the approximate information for identifying each selected star during field observations. By use of these tables, an observing list can be compiled and the stars arranged in order of increasing LST.

Once the observing list has been compiled, the astrolabe is set so there is a clear field of view for 30° around the zenith. The azimuth circle of the instrument is oriented to true north by means of a predetermined azimuth mark, by use of the magnetic needle, or by pointing the telescope tube at one of the readily identifiable stars in the observing list whose approximate azimuth has been determined. When each star in the observing list intersects the circle of constant zenith distance, the observer presses a key which records the instant of time on the chronograph.

By use of Eqs. (5–14) and (5–15), the intercept distance and azimuth angle are computed for each star observed. The same plotting procedure which was explained for Fig. 5–18 is used for the entire list. Figure 5–19 illustrates this procedure. The origin of the axes is the assumed position M of the observer. A convenient scale for the graph is determined to accommodate all the intercepts. Once all the lines of position have been plotted, a circle is drawn which is most nearly tangent to all the lines of position. This is a matter of common sense and judgment. The center of this best-fitting circle is O and represents the true position of the observer. The values of Δx and Δy in seconds of arc are then scaled

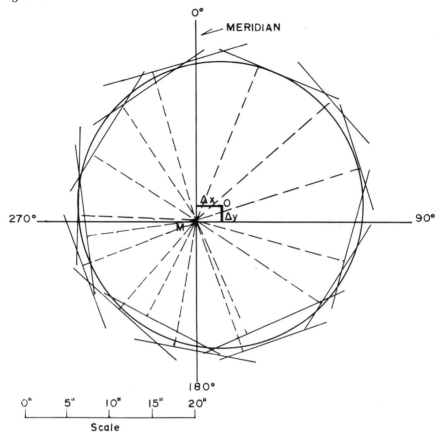

Figure 5–19. Lines of position for an entire observation list.

from the graph. The corrections in latitude and longitude then become

$$\Delta\Phi = \Delta y \tag{5-16}$$

and

$$\Delta\Lambda = \Delta x \sec \Phi. \tag{5-17}$$

These corrections are then applied properly to the assumed latitude and longitude. In the case of Fig. 5–19, the correct position O is north of the assumed position and the correction must be added. The correction in longitude must be subtracted because the actual position O is east of the assumed position. This, of course, is the case when the observer is west of Greenwich and north of the equator. For other conditions, the corrections must be applied properly.

As a matter of convenience, the value of z_0 may be assumed as any convenient value and not held rigidly to 30°. This avoids the use of negative intercepts. The physical location of each line of position is changed, but the center O remains the same. The astrolabe is a relatively simple and reliable instrument. It is light in weight and requires few adjustments. Refraction corrections need not be applied: if the observing list is selected with an equal number of stars in each quadrant, the effect of refraction is canceled. Accuracies of a few hundred feet in position may be attained. The principal weakness, in the authors' experience, is the sensitivity of the prismatic astrolabe to vibrations. When it is used on top of buildings in which air conditioners, motors, or elevators are operating, or near rail or road traffic, the mercury bath tends to tremble and the image of the reflected star is often fuzzy and may disappear entirely.

THE DETERMINATION OF AZIMUTH

The azimuth of a line is the horizontal angle which the line makes with the meridian, measured clockwise from true north. The problem of finding the azimuth of a line consists of two parts:

(1) determining the azimuth angle of the heavenly body R, and

(2) measuring the horizontal angle between the azimuth line and the heavenly body.

The problem is illustrated by Fig. 5–20. The azimuth angle $\mathcal{Z}$ of the star is measured either eastward or westward from true north. The angle K is the angle measured clockwise from the star to the line OT whose azimuth is desired. For stars east of the meridian, the astronomical azimuth A of the line OT is

$$A = \mathcal{Z} + K; \tag{5–18}$$

for stars west of the meridian, the astronomical azimuth is

$$A = 360° - \mathcal{Z} + K. \tag{5–19}$$

The first part of the problem consists of the solution of the spherical triangle formed on the celestial sphere by the pole, the zenith, and the star. The second part of the problem involves simply turning the angle between the azimuth line and the heavenly body.

At this point, a review of spherical trigonometry in Appendix A is suggested, particularly the law of sines, the law of cosines, and the relation between two angles and three sides. The solution of the azimuth problem makes use of these spherical trigonometric relationships. Figure 5–3 shows the spherical triangle $P\mathcal{Z}R$ which must be solved. Figure 5–21 shows the same triangle as viewed from the zenith. The

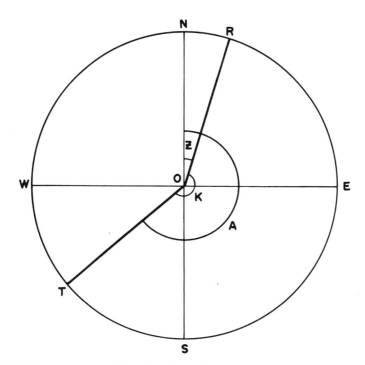

Figure 5–20. Determination of the azimuth of a line.

azimuth angle $\mathcal{Z}$ is the quantity we desire. It should not be confused with the zenith, which is designated Z.

With reference to both Figs. 5–3 and 5–21, the notation is as follows: P is the celestial pole; Z is the zenith of the observer; R is the heavenly body observed; $PZ = 90° - \Phi$; $PR = 90° - \delta$; $RZ = 90° - h$; t is the meridian angle of R; and $\mathcal{Z}$ is the azimuth angle of R, the quantity to be computed.

There are numerous methods for azimuth determination making use of the sun, moon, or stars. Three of the most common methods will be presented.

Case I. A star is observed at any hour angle, the time and altitude being measured during the observation. The following elements of the celestial triangle are known after the observation:

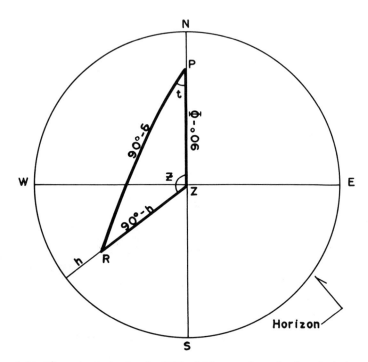

Figure 5–21. The spherical triangle PZR which must be solved to find the azimuth angle.

$$PZ = 90° - \Phi,$$
$$PR = 90° - \delta,$$
$$RZ = 90° - h,$$

and

$$t \quad = \text{RA} - \text{LST} \quad \text{(for stars east of the meridian)},$$

or

$$t \quad = \text{LST} - \text{RA} \quad \text{(for stars west of the meridian)}.$$

It is assumed that the local clock is set to local sidereal time, which is obtained from LST $=$ GST $- \Lambda$.

By the use of the spherical law of sines, we have

$$\frac{\sin t}{\sin(90° - h)} = \frac{\sin Z}{\sin(90° - \delta)},$$

or

$$\cos h \sin Z = \sin t \cos \delta, \tag{5-20}$$

from which

$$\sin Z = \frac{\sin t \cos \delta}{\cos h}.\qquad(5\text{-}21)$$

Case II. A star is observed at any hour angle, the time (but not the altitude) being recorded during the observation. The elements of the celestial triangle which are known after the observation are PZ, PR, and t.

By use of the relationship between two angles and three sides of a spherical triangle, we may write

$$\sin(90° - h) \cos Z = \sin(90° - \Phi) \cos(90° - \delta)$$
$$- \cos(90° - \Phi) \sin(90° - \delta) \cos t.$$

This equation may be simplified to

$$\cos h \cos Z = \cos \Phi \sin \delta - \sin \Phi \cos \delta \cos t.\qquad(5\text{-}22)$$

Dividing Eq. (5–20) by Eq. (5–22) produces

$$\frac{\cos h \sin Z}{\cos h \cos Z} = \frac{\sin t \cos \delta}{\cos \Phi \sin \delta - \sin \Phi \cos \delta \cos t},$$

which finally becomes

$$\tan Z = \frac{\sin t}{\cos \Phi \tan \delta - \sin \Phi \cos t}.\qquad(5\text{-}23)$$

Case III. A circumpolar star is observed at elongation. Figure 5–22, drawn in the horizon system, shows a circumpolar star at two positions R and R' at the points of eastern and western elongation about the pole P. These are points of tangency with two vertical circles ZL and ZL'; therefore, the spherical triangles PZR and PZR' have right angles at R and R', respectively. The azimuth angle is the same size whether at eastern or western elongation and is designated Z_e. It can be obtained from the law of sines,

$$\frac{\sin Z_e}{\sin(90° - \delta)} = \frac{\sin 90°}{\sin(90° - \Phi)},$$

or

$$\sin Z_e = \frac{\cos \delta}{\cos \Phi} = \cos \delta \sec \Phi.\qquad(5\text{-}24)$$

Near elongation, the azimuth of a star changes slowly. For example, the azimuth of Polaris changes less than 1 second of arc from about 4 minutes before elongation to 4 minutes after elongation. In this time period, two sets of observations can be made between the star and the azimuth mark, once with the telescope direct and once with the tele-

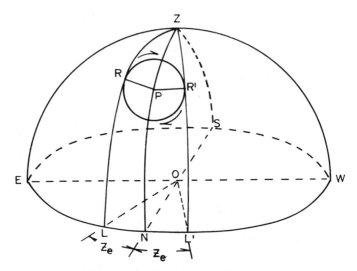

Figure 5–22. The determination of azimuth by observing a circumpolar star at elongation.

scope reversed. If Polaris is used as the circumpolar star, tables in the Almanac may be used for obtaining the azimuth at elongation. Many tedious computations can be avoided by the proper use of tables.

If the theodolite, such as the one shown in Fig. 5–23, is in proper adjustment—particularly, the horizontal axis must be truly horizontal —and good observing techniques are used, the only correction that need be made to the azimuth angle is the aberration correction. Because of the motion of the observer in the eastward direction, there is an effect of the diurnal aberration on the azimuth angle. The amount of this correction in seconds of arc is

$$\varDelta z = +0.319 \frac{\cos \varPhi \cos z}{\cos h}.\qquad(5\text{-}25)$$

When the star is east of the meridian, the correction is plus; when the star is west of the meridian, the correction is minus. Since the altitude of the pole is equal to the latitude of the observer, Eq. (5–25) may be simplified in the case of a star near the pole. In seconds of arc, the correction becomes

$$\varDelta z = \pm 0.32 \cos z.\qquad(5\text{-}26)$$

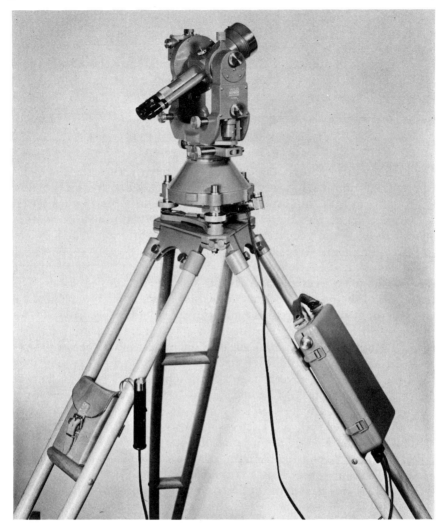

Figure 5–23. Wild T–3 Precision Theodolite (Wild Heerbrugg Instruments, Inc.).

Because the azimuth angle $\mathcal{Z}$ is reckoned either eastward or westward from the north point, whereas azimuth is measured only clockwise from north, the azimuth angle $\mathcal{Z}$ may be converted to astronomical azimuth A measured from north by these rules:

$$A = \mathcal{Z} \quad \text{(for a star east of the meridian),}$$

and

$$A = 360° - \mathcal{Z} \quad \text{(for a star west of the meridian).}$$

If an angle from the heavenly body to an azimuth mark has been measured in the procedure, its azimuth is determined by a return to Eqs. (5–18) and (5–19).

The methods outlined in Case II and Case III are those most often used in geodetic surveys. Case II should yield an accuracy of 0.′′5 of arc; Case III, 0.′′1 of arc.

RELATIONSHIP BETWEEN GEODETIC AND ASTRONOMICAL VALUES

Throughout this chapter, the authors have been very careful to point out that astronomical latitude, longitude, and azimuth are not the same as geodetic latitude, longitude, and azimuth. The capital Greek letters Φ, Λ, and A have consistently been used to distinguish the astronomical quantities from the geodetic quantities, which are designated by the corresponding small Greek letters ϕ, λ, and α.

The fact that the two are not the same is difficult for the beginning geodesist to grasp. It comes as a distinct shock to seasoned mariners who are proud of their abilities with the sextant or some other astronomical instrument. If we accept their claims of ship-positioning accuracy, or even go a little farther and grant them an error-free astronomical determination of their position, then the fact remains that the ship might be as far as $1\frac{1}{2}$ nautical miles from its true geodetic position. This fact has been alluded to several times in this book and is treated more thoroughly in Chapter 8. The mass anomalies within the earth produce corresponding gravity anomalies which deflect the plumbline away from the true vertical. This effect is known as the deflection of the vertical and has a maximum value of about 85 seconds of arc. It is called variously "deviation from the vertical", "deflection of the plumbline," and similar expressions, but the meaning is the same. Because the level bubbles on astronomical instruments are perpendicular to the plumbline, it follows that any astronomical instrument which uses level bubbles for orientation is affected by the deflection of the vertical.

The deflection of the vertical has an effect on latitude, longitude, and azimuth. The famous deflection-of-the-vertical equations relate the geodetic and the astronomical values. These equations are

$$\xi = \Phi - \phi, \tag{5–27}$$

$$\eta = (\Lambda - \lambda) \cos \phi, \tag{5–28}$$

and

$$\eta = (A - \alpha) \cot \phi, \tag{5–29}$$

where ξ is the deflection of the vertical in the plane of the meridian; Φ is the astronomical latitude; ϕ is the geodetic latitude; η is the deflection of the vertical in the plane of the prime vertical; Λ is the astronomical longitude; λ is the geodetic longitude; A is the astronomical azimuth; and α is the geodetic azimuth.

The reader should be aware of a pitfall in consulting other reference books: The symbols ξ and η are sometimes reversed. At least one textbook on geodesy reverses the above definitions of ξ and η.

The three equations (5–27), (5–28), and (5–29) are generally used in a slightly different form because the geodetic position and azimuth are normally required after the astronomical position and azimuth have been determined. Equation (5–27) can be rewritten as

$$\phi = \Phi - \xi. \tag{5-30}$$

If Eq. (5–28) is multiplied by sec ϕ, the result is

$$\eta \sec \phi = \Lambda - \lambda$$

or

$$\lambda = \Lambda - \eta \sec \phi. \tag{5-31}$$

If Eq. (5–29) is multiplied by tan ϕ, the result is

$$\eta \tan \phi = A - \alpha$$

or

$$\alpha = A - \eta \tan \phi. \tag{5-32}$$

From Eqs. (5–30), (5–31), and (5–32), it can be seen that any astronomical value can be converted to its corresponding geodetic value if we know the components of the deflection of the vertical. This is one goal of the worldwide gravity program.

Furthermore, Eqs. (5–29) and (5–28) may be equated:

$$(A - \alpha) \cot \phi = (\Lambda - \lambda) \cos \phi.$$

Multiplying by tan ϕ produces

$$A - \alpha = (\Lambda - \lambda) \sin \phi,$$

and finally

$$\alpha = A - (\Lambda - \lambda) \sin \phi. \tag{5-33}$$

Equation (5–33) is the famous Laplace equation which links differences of astronomical and geodetic longitudes with differences of astronomical and geodetic azimuths. A station where simultaneous determinations of longitude and azimuth are accomplished is known as a Laplace station.

REFERENCES

1. B. Boss, and collaborators, *General Catalogue*, Vols. I–V. Carnegie Institution of Washington, Washington, D.C.

2. Institute of Geographical Explorations, *Complete 60° Star Lists for the Position Finding by Equal Altitude Method*, Climax Stationery Co., New York.

3. International Astronomical Union, *Apparent Places of Fundamental Stars*. York House, Kingsway, London.

4. Nautical Almanac Office, U.S. Naval Observatory, *The American Ephemeris and Nautical Almanac*. Government Printing Office, Washington, D.C.

Chapter 6 Coordinate Systems

It is evident from the preceding discussions that geodesy is a branch of applied mathematics. Mathematics transforms physical observations into geodetic information of position, azimuth, elevation, distance, or the size and shape of the earth. A necessity for all calculations used, whether as an intermediate step or as an end result, is a coordinate system—the subject of this chapter. Here we must indulge in more mathematical derivations than in the preceding chapters.

The geodetic coordinate system is based on an ellipsoid of revolution which is defined by a semimajor axis, a, and a semiminor axis, b (or flattening, f), as explained in Chapter 2. We have cited positions on the earth in this system of angular coordinates of latitude and longitude, and in linear units for geoid height and elevations. Such positions may also be specified by various sets of rectangular coordinates expressed as functions of the geodetic angular and linear units as well as relative position in terms of observation units of elevation, azimuth, and range. These relationships are the first items presented. Rectangular coordinates are not to be considered an innovation but simply another tool for the geodesist. As will be seen, however, relationships between apparently dissimilar systems are readily defined with these systems.

Next, the meaning of a geodetic datum and its manner of establishment are discussed. Though the ellipsoid is the computation base, it must be fixed to the earth to form a geodetic datum, that is, a unique coordinate system. Seemingly, this topic should have come first; however, the rectangular coordinates provide a fine tool for developing this concept.

Last, methods of transforming coordinates between systems are developed. The relationship between geodetic systems, as well as that between a geodetic datum and the astronomical coordinate system, is explored. Again, rectangular coordinates facilitate this quest.

An understanding of the why and wherefore of a geodetic datum is mandatory for an understanding of geodesy.

RECTANGULAR COORDINATES

The geodetic rectangular coordinates—U, V, and W—are defined as follows:

The U axis is formed by the intersection of the geodetic meridian through Greenwich and the equatorial plane of the ellipsoid.

The V axis is perpendicular to the U axis and is in the equatorial plane. This axis is positive toward the east.

The W axis is perpendicular to the U and V axes and coincides with the minor axis of the ellipsoid.

The U, V, and W axes form a right-hand orthogonal system.

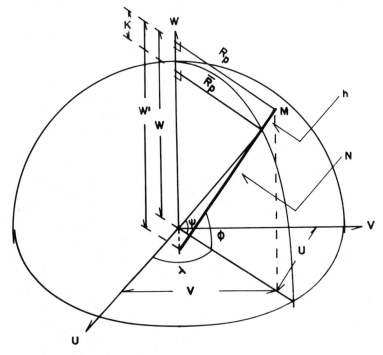

Figure 6–1. Geodetic rectangular coordinate system.

From Fig. 6–1, note that R_p and $\bar{R}_p$ are parallel to the equatorial plane, and that

$$\cos \phi = \frac{R_p}{N + h} \quad \text{and} \quad R_p = (N + h) \cos \phi,$$

where N is the radius of curvature in the prime vertical, and h the height of station above the ellipsoid (geodetic height). Also,

$$\cos \lambda = \frac{U}{R_p} \quad \text{and} \quad U = R_p \cos \lambda.$$

Substituting the expression for R_p in that for U yields

$$U = (N + h) \cos \phi \cos \lambda. \qquad (6–1)$$

Similarly, with sin $\lambda = V/R_p$ and $V = R_p \sin \lambda$,
$$V = (\mathcal{N} + h) \cos \phi \sin \lambda. \tag{6-2}$$

The procedure for W is not quite so easy. In Fig. 6–1 note that geocentric latitude, ψ, refers to the projection of point M on the ellipsoid. The geocentric and geodetic latitudes are related by Eq. (2–28a),
$$\tan \psi = (1 - e^2) \tan \phi.$$

Also, from Fig. (6–1),
$$\tan \phi = \frac{W' - K}{\bar{R}_p} \quad \text{and} \quad \tan \psi = \frac{W - K}{\bar{R}_p},$$
where $\bar{R}_p = \mathcal{N} \cos \phi$. Noting these relationships and recalling that $\tan \phi \cos \phi = \sin \phi$, we write
$$W' - K - W + K = \bar{R}_p (\tan \phi - \tan \psi),$$
$$W' - W = \mathcal{N}[\sin \phi - (1 - e^2) \sin \phi] = \mathcal{N}e^2 \sin \phi.$$

From Fig. (6–1), it is apparent that
$$W' = (\mathcal{N} + h) \sin \phi.$$

Combining the latter expression with the former yields,
$$W = [\mathcal{N}(1 - e^2) + h] \sin \phi. \tag{6-3}$$

Equations (6–1) through (6–3) present
$$U, V, \text{ and } W = f(\phi, \lambda, \mathcal{N}, \text{ and } h).$$

Now, we come to the inverse condition, that is,
$$\phi, \lambda, \text{ and } h = f(U, V, \text{ and } W).$$

Dividing Eq. (6–2) by (6–1), we have
$$\tan \lambda = \frac{V}{U}. \tag{6-4}$$

If Eq. (6–3) is divided by $R_p = (\mathcal{N} + h) \cos \phi$, and $R_p = (U^2 + V^2)^{1/2}$ is substituted in the result, an expression for ϕ may be obtained:
$$\tan \phi = \frac{(\mathcal{N} + h) W}{[\mathcal{N}(1 - e^2) + h] (U^2 + V^2)^{1/2}}. \tag{6-5}$$

For h, Eq. (6–3) is rewritten, that is,
$$h = \frac{W}{\sin\phi} - \mathcal{N}(1 - e^2). \tag{6-6}$$

As $\mathcal{N}$ and h are both functions of ϕ, Eqs. (6–5) and (6–6) must be solved by iteration. If h is (or is assumed to be) zero, Eq. (6–5) is simplified and written in a closed form [2],

$$\tan \phi = \frac{W}{(1 - e^2)\,(U^2 + V^2)^{1/2}}. \qquad (6\text{--}7)$$

Equations (6–1) through (6–3) offer an easy way to compute on the ellipsoid. For instance, suppose an equation for the length of the geocentric radius is desired. This quantity is simply equal to $(U^2 + V^2 + W^2)^{1/2}$. Also, the shortest distance between two points is the straight line distance,

$$[(U_2 - U_1)^2 + (V_2 - V_1)^2 + (W_2 - W_1)^2]^{1/2}.$$

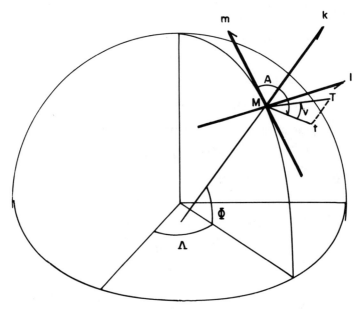

Figure 6–2. Horizon rectangular coordinate system.

Now, consider an entirely different rectangular coordinate system—the horizon coordinate system. These coordinates, m, l, and k, are expressed as functions of the observed azimuth, vertical angle, and distance (A, v, and s), respectively. Figure 6–2 illustrates the quantities of this system, the m axis being positive to the north, the l axis perpendicular to the m axis in the horizon plane, and the k axis toward the zenith, perpendicular to both the m and the l axes. M is the origin and is assumed to be a theodolite, T is the object sighted, and t is its vertical projection to the horizon plane. The spatial distance, MT, is symbolized by s. (Please note that this is not the distance along the ellipsoid.)

Equations for the coordinates of T in this system may be written simply by referring to Fig. 6–2; that is, $m = Mt \cos A$, and as $Mt = s \cos v$, then

$$m = s \cos v \cos A. \qquad (6\text{–}8)$$

Likewise

$$l = Mt \sin A = s \cos v \sin A, \qquad (6\text{–}9)$$

and

$$k = s \sin v. \qquad (6\text{–}10)$$

And, as was done with the geodetic rectangular coordinates, the inverse relationships may be obtained:

$$\tan A = \frac{l}{m}; \qquad (6\text{–}11)$$

$$s = (m^2 + l^2 + k^2)^{1/2}; \qquad (6\text{–}12)$$

$$\sin v = \frac{k}{s}. \qquad (6\text{–}13)$$

In the case of this system, the coordinates can be expressed in terms of the observation, or the observation may be stated with the coordinates.

The next step is to relate the horizon and geodetic coordinate systems.

In pursuing this relationship, it is important to remember that the horizon coordinate system is typically based on the vertical or the direction of the plumbline, not the direction of the normal to the reference ellipsoid, which is the base for the geodetic system. The horizon system is an offspring of the astronomical latitude, Φ, and longitude, Λ, not their geodetic cousins.

Obviously, Eqs. (6–1) through (6–3) can be written with astronomical coordinates and, thereby, define another U, V, and W system. With this in mind, it is apparent that the space line, MT, can have coordinate components in both systems. That is,

$$\Delta U = U_T - U_M,$$

$$\Delta V = V_T - V_M,$$

and

$$\Delta W = W_T - W_M;$$

with U, V, and $W = f(\phi$ and $\lambda)$ or $f(\Phi$ and $\Lambda)$.

Even though the astronomical coordinate may not equal the geodetic, the rectangular components in these two systems will be identical if the corresponding rectangular axes in each are parallel. This is a requirement for coordinate transformations. This mandatory condition is fixed during establishment of a datum and will be covered more fully when this topic is presented.

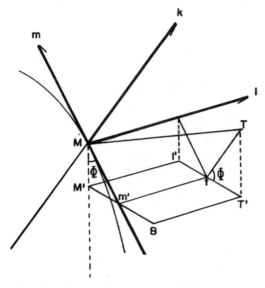

Figure 6–3. Projection of horizon system to equatorial plane.

To pursue the desired expressions relating to the horizon and geodetic systems, refer to Fig. 6–3, which is an enlargement of Fig 6–2. The plane through points M', B, and T' is parallel to the equatorial plane of the UVW system. Primed letters will indicate elements of the horizon system projected perpendicularly onto this plane. Specifically,

$$M'm' = m \sin \Phi,$$

which is the m coordinate of T projected on the M' B T' plane or, equivalently, the equatorial plane. Further, note that $tT = k$; then, with triangle tTT',

$$m'B = tT' = k \cos \Phi,$$

which is the k coordinate projected parallel to M' m'. This is always in the negative m direction for northern latitudes. In combining the projected m and k coordinates, $M'B$, these terms must be of opposite sign. From Fig. 6–3,

$$M'B = (M' m' + m' B) = - m \sin \Phi + k \cos \Phi.$$

As the l axis is parallel to the equatorial plane, these coordinates project unaltered, that is, $l = m' t = BT'$.

The projected results of the foregoing effort are shown on the equatorial plane in Fig. 6–4, from which may be seen

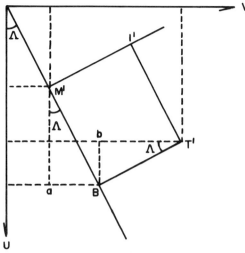

Figure 6–4. Horizon elements on the equatorial plane.

$$\Delta U = M'a - bB = M'B \cos \Lambda - BT' \sin \Lambda,$$
$$\Delta U = - m \sin \Phi \cos \Lambda + k \cos \Phi \cos \Lambda - l \sin \Lambda; \qquad (6\text{–}14)$$

and

$$\Delta V = aB + bT' = M'B \sin \Lambda + BT' \cos \Lambda,$$
$$\Delta V = - m \sin \Phi \sin \Lambda + k \cos \Phi \sin \Lambda + l \cos \Lambda. \qquad (6\text{–}15)$$

Again, by study of Fig. 6–3, ΔW is seen to be

$$\Delta W = MM' + TT'$$

or

$$\Delta W = m \cos \Phi + k \sin \Phi. \qquad (6\text{–}16)$$

Now, in Eqs. (6–14) through (6–16), the left-hand sides may be expressed with the geodetic coordinates of points M and T, and the right-hand sides in terms of the astronomical coordinates of M and observations. The significance of these equations is that they can relate the astronomical and geodetic coordinate systems.

Further, Eqs. (6–1) through (6–16) may be manipulated and combined to express observations in terms of astronomical and geodetic rectangular coordinates:

$$\tan A = \frac{\Delta U \sin \Lambda - \Delta V \cos \Lambda}{\Delta U \cos \Lambda \sin \Phi + \Delta V \sin \Lambda \sin \Phi - \Delta W \cos \Phi}; \qquad (6\text{–}17)$$

$$\sin v = \tfrac{1}{s}(\Delta V \sin \Lambda \cos \Phi + \Delta U \cos \Lambda \cos \Phi + \Delta W \sin \Phi). \qquad (6\text{–}18)$$

Or, in the same terms, the horizon coordinates:

$$m = -\Delta U \cos \Lambda \sin \Phi - \Delta V \sin \Lambda \sin \Phi + \Delta W \cos \Phi; \qquad (6\text{--}19)$$

$$l = -\Delta U \sin \Lambda + \Delta V \cos \Lambda; \qquad (6\text{--}20)$$

$$k = \Delta U \cos \Lambda \cos \Phi + \Delta V \sin \Lambda \cos \Phi + \Delta W \sin \Phi. \qquad (6\text{--}21)$$

Other combinations, including those for reciprocal observations from both ends of the line, may be derived (with care and tenacity). Line MT is fixed in space and the constant values of its ΔU, ΔV, and ΔW are independent of the horizon systems at either M or T. Remember, this is the secret of this process.

Equations (6–19) through (6–21) may be written for observations from T or M. Coordinates or observations referred to these points will bear the associated subscripts. Only the expression of the l coordinate will be cited in the following derivation.

To begin with,

$$-l_T = -\Delta U \sin \Lambda_T + \Delta V \cos \Lambda_T.$$

The negative sign preceding l_T accounts for the change of signs in ΔU and ΔV due to the reverse direction of observation. The equations for m_T and k_T are similarly written.

From Eqs. (6–14) and (6–15), the ΔU and ΔV are replaced with terms for the horizon and astronomical coordinates. The rather lengthy expression produced by this substitution is simplified by applying trigonometric identities for the sine and cosine of the difference of angles. Then l_T is expressed by

$$-l_T = l_M \cos(\Lambda_T - \Lambda_M) + m_M \sin \Phi_M \sin(\Lambda_T - \Lambda_M)$$
$$- k_M \cos \Phi_M \sin(\Lambda_T - \Lambda_M).$$

The same approach is employed to arrive at

$$-m_T, -k_T = f(l_M, m_M, k_M, \Phi_M, \Delta \Lambda).$$

The final expressions are achieved by substituting Eqs. (6–8) through (6–10) for m_M, l_M, and k_M on the right and m_T, l_T, and k_T on the left [3]. For l,

$$-\cos v_T \sin A_T = \cos v_M \sin A_M \cos(\Lambda_T - \Lambda_M)$$
$$+ \cos v_M \cos A_M \sin \Phi_M \sin(\Lambda_T - \Lambda_M)$$
$$- \sin v_M \cos \Phi_M \sin(\Lambda_T - \Lambda_M); \qquad (6\text{--}22)$$

for m,

$$-\cos v_T \cos A_T = -\cos v_M \sin A_M \sin \Phi_T \sin(\Lambda_T - \Lambda_M)$$
$$+ \cos v_M \cos A_M [\sin \Phi_T \sin \Phi_M \cos(\Lambda_T - \Lambda_M)$$
$$+ \cos \Phi_T \cos \Phi_M]$$
$$- \sin v_M [\sin \Phi_T \cos \Phi_M \cos(\Lambda_T - \Lambda_M)$$
$$- \cos \Phi_T \sin \Phi_M]; \qquad (6\text{--}23)$$

and for k,

$$-\sin v_T = \cos v_M \sin A_M \cos \Phi_T \sin(\Lambda_T - \Lambda_M)$$
$$- \cos v_M \cos A_M [\cos \Phi_T \sin \Phi_M \cos(\Lambda_T - \Lambda_M)$$
$$- \sin \Phi_T \cos \Phi_M] + \sin v_M [\cos \Phi_T \cos \Phi_M \cos(\Lambda_T - \Lambda_M)$$
$$+ \sin \Phi_T \sin \Phi_M]. \tag{6-24}$$

Symbolically, Eqs. (6–22) through (6–24) state that

$$v_T, A_T = f(v_M, A_M, \Phi_T, \Phi_M, \Delta\Lambda);$$

they express observations along a fixed line from one horizon coordinate system plus the latitude and longitude of both systems. The directions as well as the coordinates may refer to astronomical or geodetic systems at both ends of the line; or, by reversal of the sign of the left-hand term, the geodetic and astronomical systems at the same end of the line. These equations offer some interesting possibilities.

As an example, if directions are observed from both ends of a line connecting known and unknown points, the astronomical coordinates of the unknown observer can be determined. Equation (6–22) is solved for Λ_T by iteration, and then Eqs. (6–23) and (6–24) simultaneously for $\sin \Phi_T$ and $\cos \Phi_T$.

Alternatively, with coordinates (astronomical or geodetic) known at both points and observation made from point M, the directions from the other could be computed (a condition similar to the inverse case presented in Chapter 4).

Also, if the astronomical and geodetic coordinates were known at one point from which observations were made, observations referred to the geodetic system could be computed. Remember, the sign of the left-hand member would be changed and terms with a subscript T would be taken to represent geodetic elements, that is, $A_T = \alpha_M$. Of course, with geodetic coordinates known at both ends, the geodetic directions could be computed with Eqs. (6–17) and (6–18) merely by substituting geodetic coordinates.

That these equations clearly relate the astronomical and geodetic coordinate systems is their prime significance. They are also valuable as tools in deriving other working equations. However, in practical application, they suffer from dependence on measurements of vertical angles which are usually quite small and are weak owing to poor refraction corrections.

GEODETIC DATUM

We now turn our attention to the subject of the geodetic datum. The dictionary definition of datum (the singular of data) is, "something used as a basis for calculating or measuring." In geodesy, however, the word has come to have a very specific meaning: in the case at hand, this "something" is the basis for positioning points and defining the dimensions of the earth.

A geodetic datum is comprised of an ellipsoid of revolution fixed in some manner to the physical earth. An ellipsoid approximating the shape of the geoid in a limited region and having a specified relationship to a point in the region (the origin) forms a regional datum. An ellipsoid approximating the shape of the entire global geoid and having its center at the earth's center of gravity forms a global datum. An obvious requirement of any coordinate system is that it be unvarying relative to the body which is to be referred to it.

Examples of regional datums, their origins and reference ellipsoids, and reference ellipsoids for global datums are shown in Table 6–1.

TABLE 6–1

Examples of Regional and Global Datums

Regional datum	Reference ellipsoid	Origin
European Datum (ED)	International (Hayford) $a = 6,378,388$ meters $f = 1/297$	Helmert Tower, Potsdam
North American Datum – 1927 (NAD27)	Clarke 1866 $a = 6,378,206$ $f = 1/295$	Meades Ranch, Kansas
Russian Datum (Pulkovo 42)	Krasovskii 1938 $a = 6,378,245$ $f = 1/298.3$	Pulkovo Observatory, Leningrad
Tokyo Datum (TD)	Bessel 1841 $a = 6,377,397$ $f = 1/299.2$	Tokyo Observatory, Tokyo

Global datum	Reference ellipsoid
Kaula 1961	Kaula $a = 6,378,165$ meters $f = 1/298.3$
Mercury	Fischer 1960 $a = 6,378,166$ $f = 1/298.3$
Modification of the Mercury Datum 1968 (MMD 68)	Fischer 1968 $a = 6,378,150$ $f = 1/298.3$
Smithsonian Datum 1966 (SA066–C6)	SA066 $a = 6,378,165$ $f = 1/298$

Approaches to a global datum are represented by the Mercury Datum (Fischer ellipsoid) or its 1968 modification [1], the Kaula Datum used during the early 1960's by NASA [4], and the Smithsonian Datum (SA066–C6) [6].

From this discussion, a regional datum may be said to be defined by seven parameters:

(1) a, the semimajor radius of the reference ellipsoid;

(2) f, the flattening of the ellipsoid;

(3) ξ_0, the deflection of the vertical in the meridian at the datum origin, $\xi_0 = \Phi_0 - \phi_0$;

(4) η_0, the deflection of the vertical in the prime vertical at the datum origin, $\eta_0 = (\Lambda_0 - \lambda_0) \cos \Phi_0$;

(5) α_0, the geodetic azimuth from the origin along an initial line of triangulation, $\alpha_0 = A_0 - \eta_0 \tan \Phi_0$;

(6) N_0, the geoid height at the datum origin, that is, the distance between the reference ellipsoid and the geoid;

(7)   the condition that the ellipsoid semiminor axis and the earth's mean rotation axis be parallel.

The first two parameters, a and f, simply define the reference ellipsoid that approximates the shape of the regional geoid. An ill-fitting ellipsoid results in systematic departures of the ellipsoid from the geoid rather than the random departures expected because of local irregularities. For instance, if a is too large, the geoid height will systematically increase as the distance from the origin increases. Similar variation results if f is in error.

Parameters 3, 4, and 6, the deflection components and geoid height, fix the relationship between the geoid (astronomical coordinates) and the corresponding point on the ellipsoid at the origin. The origin geodetic coordinates may be assumed equal to the observed astronomical position—that is, deflection zero. Likewise, the geoid and the ellipsoid may be set tangent to one another. If these origin assumptions are grossly in error, systematic deflections and geoid height trends are noted as the survey progresses away from the origin. If the ellipsoid shape is adequate, such systematic trends are eliminated by adjusting the geodetic coordinates assumed at the origin.

It would seem that specifying the size and shape of the reference ellipsoid and fixing its angular departure and distance from the geoid at the origin would prevent it from moving with respect to the geoid. This is not true, however.

Consider that the ellipsoid may be rotated around its normal without altering any of the five conditions established thus far. Among other disadvantages, this permission to rotate precludes compliance with the condition that the earth's rotation axis and the ellipsoid semiminor axis be parallel.

This rotation is prevented through parameter 5, that is, by satisfying the azimuth equation. With a value of η_0 assigned to the origin, the astronomical azimuth, A_0, observed along a line from the origin is transformed to the corresponding geodetic azimuth, α_0, on the reference ellipsoid. It is apparent that the differences between astronomical and geodetic azimuths and longitudes at the origin are dependent; that is, we may select the difference in either longitude or azimuth arbitrarily, but not both.

Through the relationship defined by the azimuth equation at the origin, the ellipsoid is rigidly fixed to the earth.

On the basis of this initial geodetic azimuth, triangulation extends the geodetic coordinates from the datum origin. At intervals along such chains of triangles, astronomical determinations of longitude and azimuth are made. With the difference between the geodetic and astronomical longitudes and with the observed astronomical azimuths, the geodetic azimuth is computed. A comparison of this geodetic azimuth and that determined through triangle computation provides a means of adjusting the triangulation. Such triangulation points are denoted Laplace stations.

An expanded view of the origin process may be gained by returning to Eqs. (6–22) through (6–24). Recall that the basis for derivation of these equations was the equivalence of the rectangular coordinate components (ΔU, ΔV, and ΔW) of a line whether the azimuth and vertical angles are computed from geodetic coordinates or are related to the astronomical system by observation. This condition will exist provided the U, V, and W axes are parallel to their friends of the astronomical coordinate system, U_A, V_A, and W_A. With this in mind, Eqs. (6–24) and (6–22) will be modified to express an origin azimuth and vertical angle condition. As observations from one end of the line are employed, the negative sign of the left-hand term is removed. Where necessary, terms referring to the astronomical system wear a subscript A. Further, for clarity, the following substitutions are made:

$$
\begin{aligned}
v_T \leftarrow v &= v_A + \Delta v &&= \text{geodetic vertical angle} \\
v_M \leftarrow v_A &&&= \text{astronomical vertical angle} \\
A_T \leftarrow \alpha &= A + \Delta A &&= \text{geodetic azimuth} \\
A_M \leftarrow A &&&= \text{astronomical azimuth} \\
\Phi_T \leftarrow \phi &= \Phi + \Delta\Phi &&= \text{geodetic latitude} \\
\Phi_M \leftarrow \Phi &&&= \text{astronomical latitude} \\
\Lambda_T \leftarrow \lambda &= \Lambda + \Delta\Lambda &&= \text{geodetic longitude} \\
\Lambda_M \leftarrow \Lambda &&&= \text{astronomical longitude.}
\end{aligned}
$$

Substituting these relationships in Eq. (6–24) yields

$$\sin(v_A + \Delta v) = \cos v_A \sin A \cos(\Phi + \Delta\Phi) \sin \Delta\Lambda$$
$$- \cos v_A \cos A \left[\cos(\Phi + \Delta\Phi) \sin \Phi \cos \Delta\Lambda\right.$$
$$- \sin(\Phi + \Delta\Phi) \cos \Phi\left.\right]$$
$$+ \sin v_A \left[\cos(\Phi + \Delta\Phi) \cos \Phi \cos \Delta\Lambda\right.$$
$$+ \sin(\Phi + \Delta\Phi) \sin \Phi\left.\right].$$

It is assumed that the cosines of small angles—$\Delta\Phi$, $\Delta\Lambda$, and Δv—are equal to one and that the sine equals the angle; and further, that the product of $\Delta\Phi$ and $\Delta\Lambda$ is negligible. Notice the trigonometric identities for the sine and cosine of the sum of angles. Then the foregoing equation may be simplified and written

$$\Delta v = \Delta\Lambda \sin A \cos \Phi + \Delta\Phi \cos A. \tag{6–25}$$

With the same logic and tenacity, Eq. (6–22) is written,

$$\Delta A = \Delta\Lambda \sin \Phi - \frac{\tan v_A}{\cos A}(\Delta\Lambda \cos \Phi - \Delta v \sin A).$$

Then, by substituting Eq. (6–25) for Δv, we obtain

$$\Delta A = \Delta\Lambda \sin \Phi + \tan v_A (\Delta\Phi \sin A - \Delta\Lambda \cos A \cos \Phi). \tag{6–26}$$

If Eqs. (6–25) and (6–26) are satisfied at the origin, rectangular components of a line from the origin must be equal in both systems.

Additionally, with azimuth observed on or reduced to the horizon plane, Eq. (6–26) is written ($v_A = 0$) as

$$\Delta A = \alpha - A = \sin \Phi(\lambda - \Lambda),$$

which is the Laplace equation. Under these conditions, Eq. (6–25) is disregarded.

In [3] and [5], the respective authors theorized that these equations should be satisfied along more than one line emanating from the origin to insure parallelism of the axes (nonparallel axes could result from an error in $\Delta\Lambda_o$).

This requirement is in fact satisfied even though it is not included in the classical list of seven parameters specified for datum orientation.

Orientation of a global datum follows the same process except that a and f refer to an ellipsoid approximating the shape of the entire geoid and the values for ξ_0, η_0, and N_0 are considered absolute; that is, they are theoretically the variations between the geoid and the ellipsoid with its center coinciding with the earth's center of gravity. Methods of obtaining such an absolute orientation are presented in Chapters 8 and 10.

Regional datums normally are related to a global datum by specifying the difference between the regional coordinates and those on the
global datum at the regional datum origin. When all regional datums
thus can be transformed to a global datum, all land masses will have
been effectivley connected.

TRANSFORMATION OF COORDINATES

In this day and time of truly global action, it is becoming increasingly
necessary to relate distant points to the same mathematical framework.
It is impossible to apply analytical geometry to information plotted on
two unrelated graphs; it is equally difficult to compute directions and
distances between points on different geodetic systems. With this problem in mind, a discussion of transformation of geodetic coordinates
concludes this chapter on coordinate systems.

There are many techniques of performing coordinate transformation.
Some offer mathematical ease at the sacrifice of accuracy; at the other
extreme, some are accurate under all conditions but require an electronic computer for practical application. There is spirited controversy
over which method is "best" under "what" conditions.

Two methods will be presented. The first approach illustrates the

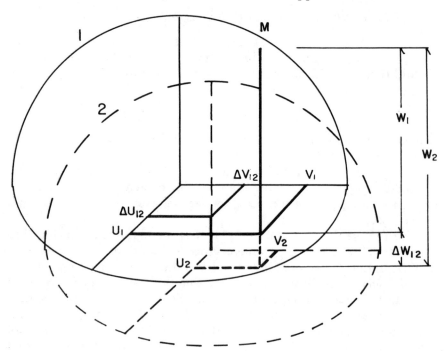

Figure 6–5. Transformation between coordinate systems.

basic transformation problem and offers a rigorous solution; the second method offers ease of computation with a slight penalty in accuracy.

Figure 6–5 illustrates a single station with coordinates in two geodetic systems—datum 1 and datum 2. With the geodetic coordinates, ϕ, λ, and h known in both systems, the U, V, and W coordinates may be computed in both systems with Eqs. (6–1) through (6–3). The differences between these rectangular coordinates are the coordinates of the center of one system in the framework of the other system. That is,

$$\Delta U_{12} = U_1 - U_2,$$
$$\Delta V_{12} = V_1 - V_2,$$

and

$$\Delta W_{12} = W_1 - W_2$$

are the coordinates of the center of system 2 in the framework of system 1.

The significance of this result is that ΔU_{12}, ΔV_{12}, and ΔW_{12} are constant for all points common to these two datums. This condition results from the axes of the two systems being parallel. Be sure to note, however, that the differences in ϕ, λ, and h for points common to both systems are not constant.

From these considerations, the steps in transforming from coordinates in datum 2 to those in datum 1 are as follows:

(1)   From a point common to both systems determine ΔU_{12}, ΔV_{12}, and ΔW_{12}.

(2)   For any other point in datum 2, compute U_2, V_2, and W_2 with Eqs. (6–1) through (6–3).

(3)   Add these rectangular coordinates to the constant values to determine the rectangular coordinates in system 1:

$$U_1 = \Delta U_{12} + U_2,$$
$$V_1 = \Delta V_{12} + V_2,$$
$$W_1 = \Delta W_{12} + W_2.$$

(4)   With these values in Eqs. (6–4), (6–5), and (6–6), compute λ_1, ϕ_1, and h_1.

Equations (6–5) and (6–6) require iteration as both sides contain terms that are functions of ϕ. An initial estimate of ϕ may be computed with Eq. (6–7). With this initial estimate entered into Eq. (6–6), a corresponding value of h is determined, which, in turn, is applied in Eq. (6–5) to compute a new value of ϕ. This process is repeated until the difference in ϕ reaches a desired limit.

This is an exact method of transformation and the accuracy is limited by errors in constant ΔU, ΔV, and ΔW and in truncation (round-off errors). To minimize errors, the constant Δ's normally are derived by considering more than one point common to both systems. Double precision is used in the electronic computation process to reduce truncation error. At least 10-place trigonometric functions are required to achieve a transformation accuracy of one millimeter [7].

Other transformation equations may be derived by differentiating Eqs. (6–1) through (6–3) or (6–4) through (6–5). Of this group, those presented by Vincenty are cited as representative of this method.

Equations (6–1) through (6–3) are differentiated with respect to ϕ, λ, h, e^2, and a. With h temporarily assumed equal to 0, three equations are achieved of the form

$$dU, dV, dW = f(N, \phi, \lambda, e^2, a, d\phi, d\lambda, dh, d(e^2), da).$$

With approximations in terms of e'^2 (square of the second eccentricity), a, and ϕ substituted for e^2 and N, the three equations are solved for $d\phi$, $d\lambda$, and dh:

$$d\phi = -[(dU \cos \lambda + dV \sin \lambda) \sin \phi - dW \cos \phi] \left(\frac{A}{\bar{a}}\right)$$
$$+ (C_1 \sin^2 \phi + C_2) \sin \phi \cos \phi; \tag{6–27}$$

$$d\lambda = -(dU \sin \lambda - dV \cos \lambda) \left(\frac{B}{\bar{a}}\right) \sec \phi; \tag{6–28}$$

$$dh = (dU \cos \lambda + dV \sin \lambda) \cos \phi + dW \sin \phi$$
$$+ C_3 \sin^2 \phi + C_4 \sin^4 \phi - da; \tag{6–29}$$

where

$$dU = U_1 - U_2,$$
$$dV = V_1 - V_2,$$
$$dW = W_1 - W_2,$$
$$d\phi = \phi_1 - \phi_2,$$
$$d\lambda = \lambda_1 - \lambda_2,$$
$$dh = h_1 - h_2,$$
$$d(e^2) = e_1^2 - e_2^2,$$
$$da = a_1 - a_2,$$
$$\bar{a} = \tfrac{1}{2}(a_1 + a_2),$$
$$\bar{e}'^2 = \tfrac{1}{2}(e_1'^2 + e_2'^2),$$

($\bar{a}$ and $\bar{e}'^2$ being mean values of the old and new ellipsoids);

$$C_1 = -\tfrac{1}{2}\bar{e}'^2 d(e^2),$$
$$C_2 = \frac{\bar{e}'^2}{\bar{a}} da + (1 + \bar{e}'^2) d(e^2),$$
$$C_3 = \tfrac{1}{2}\bar{e}'^2 da + \tfrac{1}{2}\bar{a} d(e^2),$$
$$C_4 = \tfrac{1}{2}\bar{e}'^2 (\bar{e}'^2 da + \tfrac{1}{2}\bar{a} d(e^2)),$$
$$A = 1 + \bar{e}'^2 - \tfrac{3}{2}\bar{e}'^2 \sin^2 \phi,$$
$$B = 1 - \tfrac{1}{2}\bar{e}'^2 \sin^2 \phi.$$

The symbols cited in [7] have been altered to prevent conflict with conventions established by the authors.

Now to make amends for the assumption that $h = 0$, multiply $d\phi$ and $d\lambda$ by $1 - h/R$, where R is an approximate value of the earth's radius, 6370 km, and h is the geodetic height of system 1. Obviously, to convert to arc seconds, the results must be divided by the sin $1''$.

Equations (6–27) through (6–29) offer a ready means of transforming coordinates between systems with a desk calculator. They have been devised to eliminate the need for special tables. Between any two geodetic coordinate systems the constant values of $\bar{a}$, $\bar{e}'^2$, C, A, and B need be computed once for coordinate conversion of any number of points.

At the origin or other point where dU, dV, and dW were determined, this transformation is exact. The accuracy decreases slowly with distance from the origin, and depends on the magnitude of dU, dV, dW, $d(e^2)$, and da. According to [7], it seems probable that errors will be within 1 part in 10^8 for all cases of practical significance.

REFERENCES

1. Irene Fischer, Mary Slutsky, F. Ray Shirley, and Philip Y. Wyatt III, New pieces in the picture puzzle of an astrogeodetic geoid map, paper presented to the XIVth General Assembly of the International Union of Geodesy and Geophysics, Lucerne, Switzerland, September, 1967.
2. R. A. Hirvonen, The reformation of geodesy, *J. Geophys. Res.* **66**, No. 5, (1961).
3. Martin Hotine, A Primer of Non-Classical Geodesy, A report of Study Group No. 1, I.G.A. Toronto, 1957.
4. W. M. Kaula, A Geoid and World Geodetic System Based on a Combination of Gravimetric, Astrogeodetic and Satellite Data, NASA TN 702, NASA, Washington, D.C., May 1961.
5. M. S. Molodenskii, *et al*, *Method for Study of the External Gravitational Field and Figure of the Earth*, The Israel Program For Scientific Translation with The National Science Foundation, Washington, D.C., and the Department of Commerce, USA, 1962 (Translated from Russian).
6. Smithsonian Astrophysical Observatory (edited by C. A. Lindquist and G. Veis), Geodetic Parameters For a 1966 Smithsonian Institution Standard Earth, Spec. Rept. 200.
7. T. Vincenty, Transformation of geodetic data between reference ellipsoids, *J. Geophys. Res.* **61**, No. 10 (1966).

Chapter 7 *Electronic Surveying*

Electronic measurement techniques have almost revolutionized the science of geodesy. Beginning in the years immediately following World War II, the electronic methods have slowly but surely developed to the point where they are now commonly accepted. With continued development, they may some day become conventional rather than specialized. Except for small tasks associated with plane surveying, the distance-measuring methods using tapes, stadia, subtense bars, and similar equipment may become a rarity. The growth of missile test ranges throughout the world has further stimulated the use of electronic surveying equipment of all types.

The universal acceptance of electronic surveying methods has given the geodesist a new tool with which to tie together continents and isolated islands. It has enabled him to establish by rapid and economical means the control network for photogrammetric and cartographic projects. So vast is the whole comprehensive picture of electronic surveying that it deserves a separate textbook and reference book, recently supplied by Laurila [11], which is highly recommended for those readers who wish to dig deeper into the fundamentals and applications of electronic surveying.

In this introductory book, covering the entire general field of geodesy, the authors present sufficient material for the reader to understand the basic principles of electronic mensuration, and for him to acquire an appreciation for the capabilities and limitations of the many systems in current use.

Historically, World War II is the dividing line between the research and the development of electronic surveying systems. The basic research goes back to the nineteenth century when Maxwell of England proved that the propagation velocity of electromagnetic radiation is the same as that of light. Hertz of Germany proved experimentally that radio waves were reflected from solid objects in the same manner as light rays. In 1904, a German engineer named Hulsmayer was able to detect and track distant objects. The inventor of the wireless, Marconi of Italy, recommended the use of Hulsmayer's invention in navigation, and the use of short waves for radio detection. Appleton and Barnett of England measured the height of the ionosphere in 1924 by use of a frequency-shift continuous-wave method. The following year, Breit and Tuve of the United States accomplished the same feat by transmitting a train of very short pulses of electromagnetic energy and determining the elapsed time until the reflections were received. In 1928, Everitt of the United States invented the radio altimeter for measuring the distance from an aircraft to the ground. The first pulse-type equipment using cathode-ray tubes as indicators was built by the Scottish physicist, Sir Robert Watson-Watt, in 1935 in England.

This was the heritage of research that was available to development engineers at the outbreak of World War II. Spurred onward by the wartime need for navigation and blind-bombing equipment, five years and nearly three billion dollars later the result was a variety of electronic measurement devices which have since been used for surveying purposes. So dramatic was the development of the electronic systems that by early 1946 the U.S. Army Signal Corps used reflected radar signals to measure the distance to the moon.

These first-generation electronic surveying systems have been used, modified, improved, and supplanted by later developments as the state of the art has advanced. Altogether there are about a dozen different systems from which an electronic surveyor may choose, depending on the mission to be accomplished.

Whereas the conventional geodesist measures one baseline, then turns a series of angles, the electronic surveyor is not at all concerned with angles. Using electromagnetic energy, he measures either distances, or distance differences, or distance sums. The absence of angle measurements and the complete reliance on distance measurements of some type have led to the coinage of a new word, "trilateration," to replace "triangulation," wherein angles are measured.

From a geometrical and functional standpoint, electronic surveying instruments can be grouped into six categories:

(1) circular systems, which measure distances directly,

(2) hyperbolic systems, which measure distance differences,

(3) circular-hyperbolic systems, which combine categories 1 and 2,

(4) hyperbolic-elliptical systems, which use both the sum and the difference of distances, the sums producing a family of ellipses and the differences producing a family of hyperbolas,

(5) baseline measuring systems, and

(6) altitude measuring systems.

CIRCULAR SYSTEMS

In circular location methods, a point to be located is determined by the intersection of two distance-circles in space or on the reference ellipsoid.

In Fig. 7–1, the two ground stations A and B are located at known geographic positions which for the purpose of illustration, are 121.164 miles apart. If the mobile station, either on land, on sea, or in the air, is 82.723 miles from Station A and 114.675 miles from Station B, it must be at either point M or M', the points of intersection of the two distance-circles. It must be assumed that the observer at the mobile station has sufficient knowledge of his general location to enable him to identify which of the two positions is the correct one.

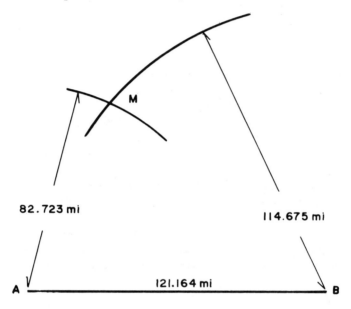

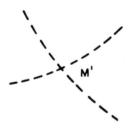

Figure 7–1. Circular location.

These two circles are obtained by direct electronic measurements of the distances between each of two fixed ground stations and the airborne (or surface) mobile station. The echo, or pulse, principle is used in most of the circular systems. A short pulse of electromagnetic energy is transmitted at one station, received at a distant station, then transmitted back to the original station. The time interval, T, required for the round trip, minus fixed delays in the equipment, is multiplied by the velocity of electromagnetic energy, V, and divided by 2 to get the distance between the two stations, or stated mathematically,

$$D = \frac{V \cdot T}{2}. \tag{7-1}$$

This gives one of the two distance-circles. The same procedure is repeated with another distant station to furnish the second distance-circle.

An analogy to the echo, or pulse, principle can be found in the case of a thirsty prospector stranded on the desert without water because his canteen slipped off his burro's pack somewhere back in the sagebrush. In the distance are the Sierra Nevadas rising sharply from the desert floor. He has *a priori* knowledge that a sparkling mountain brook flows at the foot of the mountains. Whether or not he can make it there before his strength gives out is the question. From his high school physics course he vaguely recalls that sound travels through air at sea level at a velocity of approximately 1080 feet per second. With one eye on his Accutron, he blasts away with his Colt .44. Sometime later he hears the echo and notes the time. Simply by multiplying the velocity by the time interval and dividing by two he determines the distance to the mountain. He then goes through the agonizing appraisal of whether he can make it to the mountain, or if sometime later another prospector on the quest for gold will stumble over a pile of bleached bones.

Ordinary clocks are not accurate enough to measure extremely small time intervals. For instance, to achieve an accuracy of 10 meters in a distance measurement requires a timing accuracy of 0.066 microsecond. To obtain an accuracy of one meter requires 0.0066 microsecond. Therefore the cathode ray tube (CRT) has replaced the clock for the measurement of these small time intervals. A beam of electrons, sharply focused on the fluorescent screen of the CRT, is swept across that screen by an alternating magnetic field. The time interval to be determined is measured with the aid of the beam of electrons which gives a visual indication and which can be turned on and off by the transmission and reception of the pulse of electromagnetic energy. The visual indication, which is a bright line or time base on the fluorescent screen of the CRT, can be translated, by a scale-factor and the length of the time base, into the time interval between transmission and reception. This time interval is a function of the measured distance.

Because the measured distance is a function of the velocity of electromagnetic energy, it is necessary to have a precise knowledge of the velocity of propagation if the final result is to be acceptable. Through the years, many determinations of velocity have been made with various methods. The principal ones are shown in Table 7–1 [2]. The 1941 Anderson value was used in the earlier electronic surveying work but later experience and investigation revealed it to be low by about 16 or 17 kilometers per second. In 1957, the International Union of Geodesy and Geophysics adopted the value of 299,792.5 kilometers per second for the velocity *in vacuo*, henceforth designated V_0. This value was agreed upon for use in electronic surveying.

There are several different systems available which use the circular method. Shoran, and its later version called Hiran, is by far the most important for geodesy because of its globe-girdling applications. For this reason, and because all circular systems are quite similar, Shoran will be explained in detail whereas the other circular systems will be treated briefly.

Shoran and Hiran. Shoran, a word derived from *sho*rt *ra*nge *n*avigation, was developed by the Radio Corporation of America just prior to World War II. Originally designed as a navigation system, Shoran gave early promise as an all-weather bombing system. Even before the end of World War II, Captain Carl I. Aslakson of the U.S. Coast and Geodetic Survey saw the possibility of applying Shoran to geodetic surveys. During the following years he devoted his talents and enthusiasm to its development and use, bringing the system to its enviable position in electronic surveying.

By 1949, improvements and modifications to Shoran prompted the coinage of a new name, "Hiran," which is derived from the words *hi*gh precision Sho*ran*. Very often Hiran has been erroneously considered to be a separate and distinct system. However, the principles are the same and the operating frequencies are identical. Modifications were made to handle the problem of the signal-strength or "rise-time" error; a special gain riding system is employed at both ground and airborne stations; the method of making the compensation for the nonlinearity error is different; the dials read to one more decimal part of a mile; and various minor changes were made to improve and upgrade Shoran. Furthermore, the numerous governmental and commercial survey agencies using Shoran made modifications to suit their particular need. Some agencies have retained the original ground station and modified the airborne station only. For these reasons, there is not a clear line of demarcation between Shoran and Hiran. Because the basic principles

TABLE 7–1

The Velocity of Light

Year	Observer	Method	Velocity in vacuo (km/sec)
1857	Weber and Kohlrausch	Ratio between electromagnetic and electrostatic units	310,800
1868	Maxwell	Ratio between electromagnetic and electrostatic units	284,300
1879	Michelson	Rotating mirror	299,910
1882	Newcomb	Rotating mirror	299,860
1882	Michelson	Rotating mirror	299,850
1902	Perrotin	Toothed wheel	299,880
1906	Rosa and Dorsey	Ratio between electromagnetic and electrostatic units	299,790
1923	Mercier	Guided waves	299,782
1924	Michelson	Rotating mirror	299,802
1926	Michelson	Rotating mirror	299,798
1928	Karolus and Mittelstadt	Kerr cell	299,786
1929	Birge	Statistical estimate	299,796
1934	Birge	Statistical estimate	299,776
1935	Michelson, Pease and Pearson	Rotating mirror	299,774
1937	Anderson	Kerr cell	299,771
1940	Huttel	Kerr cell	299,771
1941	Anderson	Kerr cell	299,776
1941	Birge	Statistical estimate	299,776
1944	Dorsey	Comprehensive analysis	299,773
1947	Essen	Cavity resonator	299,793
1947	Smith, Franklin and Whiting	Gee-H	299,786
1947	Jones and Cornford	Oboe	299,782
1948	Essen	Cavity resonater	299,792
1948	Bergstrand	Kerr cell – photo cell	299,796
1949	Aslakson	Shoran	299,792.4
1950	Essen	Cavity resonator	299,792.5
1950	Bol	Cavity resonator	299,789.3
1950	Bergstrand	Geodimeter	299,792.7
1951	Bergstrand	Geodimeter	299,793.1
1952	Froome	Microwave interferometer	299,792.6
1953	Froome	Microwave interferometer	299,793.0
1953	Ridgeway and Caithness Bases, England	Field geodetic project	299,792.3
1954	National Mapping Office, Australia	Field geodetic project	299,792.35
1955	Oland Bale Line, Sweden	Field geodetic project	299,792.4
1958	Froome	Microwave interferometer	299,792.50
1958	Velichko, Vasilyev, Khomaza, and Bolshakovat	Russian copy of the Geodimeter	299,792.7

are the same, the more descriptive name Shoran will be used throughout this book and will be construed to embrace Hiran as well.

Shoran is an electronic measuring system that indicates distances
from an aircraft (or surface craft) to each of two fixed ground stations.
In operation, the airborne station transmits radio signals alternately to
each of two ground stations. These signals are received and retransmitted back along the same ray path. By automatic measurement of the
time required for the signals to traverse each round-trip path, the airborne equipment obtains an accurate value for the distances to the
ground stations. This information, together with knowledge of the
flying height, the elevation of the ground station, and certain data regarding atmospheric conditions, permits very accurate positioning of
the aircraft relative to the two ground stations at the instant of each
reading. In areas where the geodetic positions of the ground stations
are known, the aircraft position may be computed.

The distance of the Shoran aircraft from each ground station is
determined by means of the "echo-timing principle." Echo timing
measures the time required for a short train of radio energy called a
pulse to travel over a distance and back to the starting point. Unlike the
usual airborne radar system, which relies on the reflection of the transmitted pulse from some object, Shoran uses a ground station which, in
itself, is both a radio receiver and transmitter. The pulse is received by
the ground station, greatly amplified, then retransmitted to the airborne station. The radio echo of this pulse on its return from the ground
station is made to produce a "pip" on the screen of a cathode ray tube.
If, at the time the pulse is transmitted by the airborne station, a marker
pip is made simultaneously on the screen, the distance between the
marker pip and the echo pip on the screen is a direct indication of the
time which elapsed between the formation of the first pip and the formation of the second one. Since the velocity of electromagnetic energy
is known, the distance between pips is also an indication of the round-
trip ray path distance between the airborne and the ground station.
Alternately, the airborne station transmits on a different frequency to a
second ground station and the same process is repeated so that a second
pip is displayed on the CRT. One of the ground stations is identified as
the "drift" station and the other called the "rate" station. The manner
in which the marker pip, the drift pip, and the rate pip are displayed on
the CRT is shown in Fig. 7–2.

In practice, a "pulse advance system" is used to produce transmitted pulses sufficiently earlier than the corresponding marker pulses
so that a signal returning from its round trip arrives just in time to
meet the corresponding pulse at the CRT. The amount of pulse advance is variable and is calibrated so that the distance from the ground

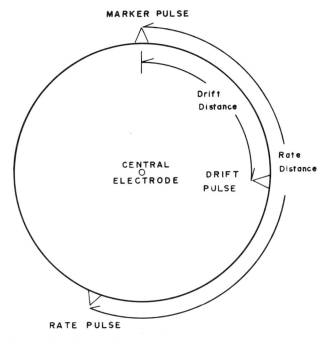

Figure 7–2. Shoran circular time base.

station is read directly. During operation, the device is constantly adjusted by the airborne operator until the leading edge of the echo pip
coincides with that of the marker pip. The distance from the aircraft to
the ground station can then be read in miles and three decimal parts

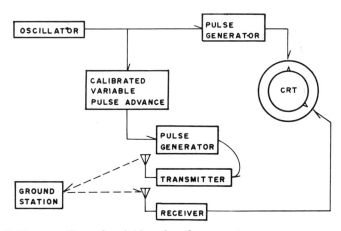

Figure 7–3. Shoran calibrated variable pulse advance system.

thereof on a dial scale. To make possible a fix of the aircraft position, a second pulse advance system is used to display the distance to the other ground station. Figure 7–3 is a schematic drawing showing the general principles of the system.

The alternate airborne transmissions to the two ground stations are made at 0.05-second intervals through the use of a comutator. Typical airborne transmission frequencies are 210 and 230 megahertz (MHz). The ground stations both respond on a common frequency, with 310 MHz being commonly used. The coverage of the airborne station is 360° in azimuth and ± 45° in elevation. The coverage of the ground station is 90° in azimuth and ± 45° in elevation. Components of airborne and ground stations are shown in Figs. 7–4, 7–5, and 7–6.

Shoran has three applications in electronic surveying: photogram-

Figure 7–4. Shoran ground station (official U.S. Air Force photograph).

Figure 7–5. Hiran ground station antenna (official U.S. Air Force photograph).

metric mapping, hydrographic surveying, and geodetic trilateration. In photogrammetric mapping, the aircraft is equipped with the Shoran airborne station and an aerial camera of a type suitable for the specific mission to be performed. The flight path is planned to provide the desired coverage, while the intervalometer setting is computed to give the necessary overlap between consecutive photographs. The timing system

Figure 7–6. Hiran airborne station and ancillary equipment intalled in an
RC–130A aircraft (official U.S. Air Force photograph).

aboard the aircraft is arranged so that an exposure is made by the aerial
camera simultaneously with an exposure by a 35-mm recording camera
mounted in a position to photograph the distance dials of the Shoran set.
The two Shoran distance measurements, together with the flight altitude of the aircraft obtained by use of a radio altimeter or Airborne
Profile Recorder, furnish the information necessary for establishing the
aircraft's spatial position at the instant the aerial camera is triggered.
Figure 7–7 shows in plan view the geometry for one single aerial photograph. The rate and drift distances from the aircraft to the rate and drift
stations, respectively, are recorded. Similar measurements are recorded
for each aerial photograph in the flight line. Parallel strips are flown
with suitable side overlap if a mosaic is desired. For each exposure, the
corresponding ground nadir is determined later in the laboratory.
Accuracies of $\pm$ 30 feet are common in Shoran-controlled photography.

For low-altitude photography, the flight lines can be flown by a
competent pilot and Shoran operator team, particularly when adequate
flight maps are available. The problem is not so easy when high altitude
missions are flown. For this type of photogrammetric mapping, a
Straight Line Indicator, shown in Fig. 7–8, is often used. It is essentially

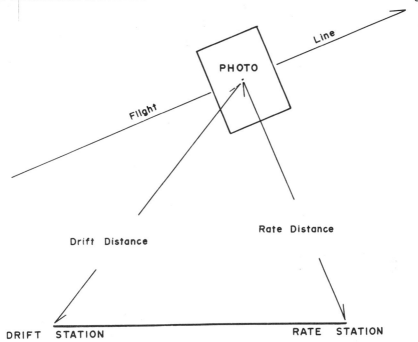

Figure 7–7. Shoran-controlled aerial photography.

a miniature of the entire Shoran system: two ground stations and the airborne station. A base plate, which represents the earth's surface to a scale of 1 : 1,000,000 to permit the use of a standard aeronautical chart as an overlay, covers an area 300 miles wide and 400 miles long on the earth. The two "ground stations" are established in their proper location on the base plate. Each ground station contains a mileage counter, a servo system, and a gear which drives a threaded rod. The two threaded rods represent the measured Shoran ray path distance from the airborne station to the two ground stations. The threaded rods connect to a moving block assembly which represents the position of the aircraft. The moving block is driven along the chart by the rotation of the threaded rods. Its trace represents the flight path as actually flown. This actual flight path, when compared with the predetermined flight path, gives the right or left error indication. Either by voice from the Shoran operator to the pilot over the intercom, or by use of a Pilot Directional Indicator (PDI), the pilot receives information regarding course corrections. The PDI is actuated by signals from a wiper arm and potentiometer on the moving block assembly. By use of the Straight Line Indicator and the Pilot Directional Indicator, flight lines can be flown with a standard error of about 200 meters [11].

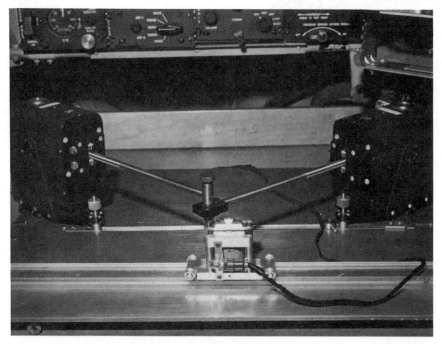

Figure 7–8. Shoran Straight Line Indicator (official U.S. Air Force photograph).

Another use of Shoran is in hydrographic surveying. One notable project was the work carried out in the area of the Aleutian Islands in 1945 by the U.S. Coast and Geodetic Survey. In less than one week, 720 kilometers of sounding lines were run. Shoran was used extensively in the early days of off-shore oil exploration in the Gulf of Mexico, although it has given way generally to hyperbolic systems. The same principles that were mentioned in airborne photogrammetric mapping also apply to sea operations. The procedures are simpler because the craft is at sea level and traveling at a slow speed.

By far the most important application of Shoran is in the establishment of geodetric trilateration networks for the purpose of making ties between continents, bringing remote land areas and isolated islands into the same survey system, and providing control for aerial mapping previously discussed. In the geodetic operation, the aircraft flies across the line whose measured length is desired. A Shoran ground station is placed at each end of the line. Figure 7–9 shows the situation in both plan and profile views. The simultaneous slant ranges to each of the two ground stations are recorded by a 35-mm motion picture camera which photographs the airborne Shoran instrument panel at regular intervals of time, usually every two seconds. At some point in crossing the line,

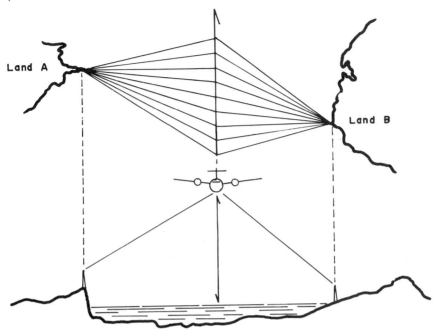

Figure 7–9. Shoran line-crossing procedure.

the sum of the two distances will reach a minimum. This condition occurs when a vertical plane through the airborne station also contains both ground stations. Figure 7–10 shows a graph of the minimum sum distance versus the recorder exposure number. Twelve line crossings are normally made, six at each of two predetermined altitudes. Geometric and velocity corrections are applied to the mean value of the measured distances to reduce it to the ellipsoidal distance between the two ground stations. This reduced distance becomes the working value for use in further computations and adjustments in the trilateration network. Each survey agency has its preferred reduction equations which are generally converted to graphs for use by the technicians in their computations.

Because the reduction of the ray path distance to the ellipsoidal distance is fundamental to all airborne electronic surveying, a derivation of the velocity and geometric corrections is presented. First, the velocity correction is explained. Shoran equipment, as well as other electronic measuring equipment, is designed for a constant velocity of propagation. This value is fixed in the instrument by a crystal of a given frequency. Because radio waves travel through air of varying pressure, temperature, and moisture content, the velocity changes along the ray path. It is necessary, therefore, to correct each measured distance by an

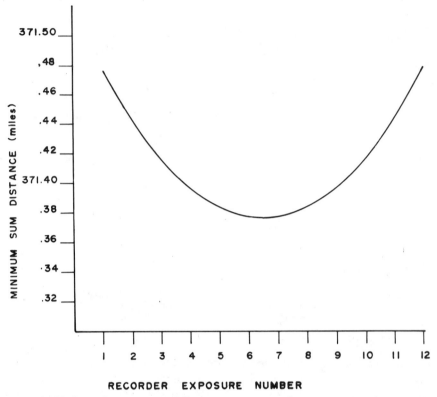

Figure 7–10. Sum of two measured distances versus recorder exposure number.

amount which is a function of the difference between the assumed conditions and the actual conditions. The lapse rates of pressure, temperature, and relative humidity along the ray path may be determined in several ways. The best method is to collect meteorological data along the ray path by use of a specially equipped aircraft.

Although the velocity of radio waves *in vacuo* is constant and equal to that of light, in the atmosphere the effective propagation is a function of the atmosphere through which the ray passes. The velocity of propagation in the atmosphere is

$$V = \frac{V_0}{n}, \tag{7–2}$$

when V is the velocity in the atmosphere, V_0 is the velocity *in vacuo*, and n is the refractive index.

There are a variety of relationships in use which express the variation of the index of refraction with changes in pressure, temperature and relative humidity. These expressions generally take the following form with small differences in the constants [1]:

$$(n-1)\ 10^6 = \frac{79.5\ (P-e)}{T} + 3.80 \times 10^5 \frac{e}{T^2}, \qquad (7\text{--}3)$$

in which P is the total pressure in millibars, e is the partial vapor pressure in millibars, and T is the absolute temperature, or $273°$ plus centigrade temperature.

Once an average value for n is established, the average velocity in the atmosphere is computed by a variation of Eq. (7–2):

$$\bar{V} = \frac{V_0}{\bar{n}}, \qquad (7\text{--}4)$$

where $\bar{V}$ is the mean propagation velocity along the ray path, V_0 is the velocity *in vacuo*, and $\bar{n}$ is the average refractive index.

From the Shoran distance dial reading, the true Shoran ray path distance is computed:

$$D_A = \frac{\bar{V} \cdot D_d}{f_t}, \qquad (7\text{--}5)$$

where D_A is the Shoran ray path distance in statute miles, $\bar{V}$ is the mean propagation velocity along the ray path in statute miles per second, D_d is the Shoran dial reading in statute miles, and f_t is the design Shoran timing frequency in cycles per second.

With the value of the Shoran arc established by use of Eq. (7–5), the geometric correction must now be made to reduce the arc to the ellipsoidal distance. Figure 7–11 shows the ground antenna located at point P_1 a distance K above point M_1 which is on the reference ellipsoid. The aircraft is at P_2, a distance H above M_2 which is on the reference ellipsoid. The geocenter is designated O, while R_α is the radius of curvature of the ellipsoid along the normal section containing points M_1 and M_2. The value of R_α is given by Eq. (2–23).

The first step is to reduce the ray path distance, or the arc D_A, to its corresponding chord distance D_C. The arc and chord are removed from Fig. 7–11 and shown in Fig. 7–12, with added construction to show the average radius of curvature R_1 of the arc D_A. The point O_1 is the center of the circle of which the arc D_A is a segment. The center angle of the arc D_A is designated ϕ and may be expressed as a function of D_A and the radius R_1, or

$$\phi = \frac{D_A}{R_1}. \qquad (7\text{--}6)$$

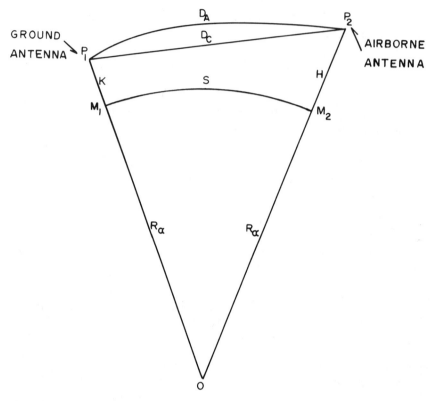

Figure 7–11. Geometry of the Shoran reduction problem.

From the triangle $O_1 P_1 P_2$,

$$D_C = 2R_1 \sin \frac{\phi}{2}. \tag{7–7}$$

By substituting Eq. (7–6) into Eq. (7–7),

$$D_C = 2R_1 \sin \frac{D_A}{2R_1}. \tag{7–8}$$

If $\sin (D_A/2R_1)$ is expressed as a series, the result is,

$$\sin \frac{D_A}{2R_1} = \frac{D_A}{2R_1} - \frac{D_A^3}{48R_1^3} + \ldots \ldots \tag{7–9}$$

If only the first two terms of the development are used, Eq. (7–8) may be rewritten as

$$D_C = 2R_1 \left(\frac{D_A}{2R_1} - \frac{D_A^3}{48R_1^3} \right). \tag{7–10}$$

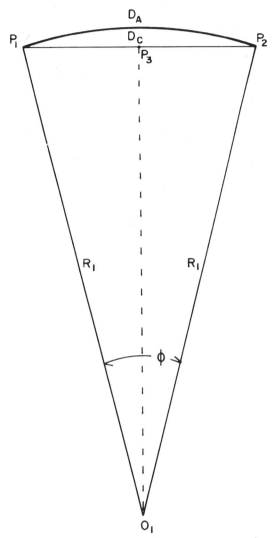

Figure 7–12. Shoran ray path, chord distance, and radius of curvature.

The chord distance finally becomes

$$D_C = D_A - \frac{D_A^3}{24R_1^2}. \tag{7–11}$$

The value of R_1 to be used in Eq. (7–11) depends upon the accuracy desired. For the average radius of curvature of the Shoran ray path, the following equation can be used for most purposes when the flying altitude is 5 miles or less and the distance does not exceed 250 miles:

$$R_1 = r(2 + 1.9h), \tag{7-12}$$

where r is the average radius of curvature of the earth and is equal to 3959 statute miles, and h is the average altitude of the ray path in statute miles.

For distances and flying altitudes greater than those specified above more rigorous derivations of the radius of curvature of the ray path must be made from observed meteorological data. These data are collected variously on the ground, from balloons, by the aircraft flying the Shoran mission or by an aircraft whose prime function is to gather meteorological data. The amount of data needed depends upon the accuracy specified for the survey and the methods used by the survey agency performing the mission.

The next step in the geometric correction is that of reducing the value of the chord distance D_C onto the ellipsoid. In Fig. 7–13, the

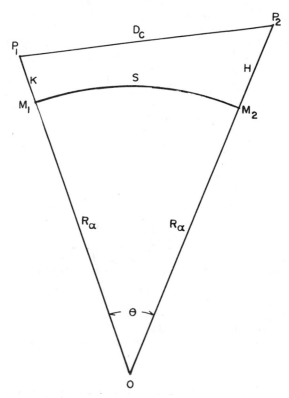

Figure 7–13. Reduction of Shoran chord distance onto the ellipsoid.

distance S is desired, with D_C already known by means of Eq. (7–11). The angle θ is the geocentric angle, and all other quantities are as defined earlier for Fig. 7–11.

Using the law of cosines, and referring to Fig. 7–13,

$$D_C^2 = (R_\alpha + K)^2 + (R_\alpha + H)^2 - 2(R_\alpha + K)(R_\alpha + H)\cos\theta \qquad (7\text{–}13)$$

from which

$$\cos\theta = \frac{(R_\alpha + K)^2 + (R_\alpha + H)^2 - D_C^2}{2(R_\alpha + K)(R_\alpha + H)}, \qquad (7\text{–}14)$$

or

$$\theta = \arccos\left[\frac{(R_\alpha + K)^2 + (R_\alpha + H)^2 - D_C^2}{2(R_\alpha + K)(R_\alpha + H)}\right]. \qquad (7\text{–}15)$$

Also, from Fig. 7–13,

$$S = R_\alpha \cdot \theta,$$

so that

$$S = R_\alpha \arccos\left[\frac{(R_\alpha + K)^2 + (R_\alpha + H)^2 - D_C^2}{2(R_\alpha + K)(R_\alpha + H)}\right]. \qquad (7\text{–}16)$$

The Shoran ray path is a straight line in a vacuum but is refracted downward when traveling through the atmosphere. Bending of the path is greatest near the ground but approaches a straight line at higher altitudes where the atmospheric pressure and moisture content become less. The maximum range of a Shoran aircraft from a ground station is thus dependent upon the curvature of the ray due to the refraction of the atmosphere, the altitude of the aircraft, the elevation of the ground station, and the elevation and location of any obstacle between the aircraft and the ground station. This is clearly shown in Fig. 7–14. Graphs are generally used in operating units for flight planing. Table 7–2 shows the maximum distance which can be measured between a ground station and an airborne station in the special case of the ground station located at sea level with no obstructions between it and the aircraft. The table shows the limiting distance for each 5000

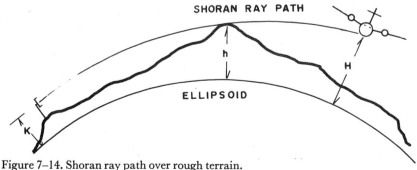

Figure 7–14. Shoran ray path over rough terrain.

feet of flying altitude up to 40,000 feet. The aircraft must, of course, be within line of sight of both ground stations. In the case of line crossings, the line-of-sight distances are additive because the aircraft is between the two ground stations and measures in both directions. The maximum length of a line successfully measured and used in a trilateration network is approximately 576 miles.

TABLE 7–2

Line-of-Sight Distance versus Altitude

Aircraft Altitude (feet)	Aircraft Distance (statute miles)
5,000	96
10,000	136
15,000	167
20,000	192
25,000	215
30,000	236
40,000	272
50,000	304
60,000	333

In the late 1940's, Shoran was used on rather modest jobs. In the early 1950's, the survey of the Atlantic Missile Range was begun. Figure 7–15 shows the trilateration network extending from Florida through Puerto Rico. Figure 7–16 shows its continuation through Trinidad and the conventional triangulation tie into South America. The probable error of the geographic position of a Shoran trilateration station relative to another station in the network is slightly less than two meters.

In 1952, Shoran was used to make the tie across the Mediterranean between the European Datum and that of Africa to complete the long north-south arc measurement which was described in Chapter 1 and shown by Fig. 1–2. Three ground stations were established in Crete on the European Datum; three were on the African side, with two in Egypt and one in Libya near the Egyptian border. This tie enabled Hough to use a single connected arc of 7000 miles in his solution of the earth's dimensions.

Soon after the Mediterranean crossing, the North Atlantic tie was begun by the 1370th Photo-Mapping Wing. There had long been a need for a tie between the European Datum and the North American Datum. The distance between the two is greater than can be spanned by either conventional or flare triangulation and this was, of course, before the days of geodetic satellites. The best-laid plans of solar

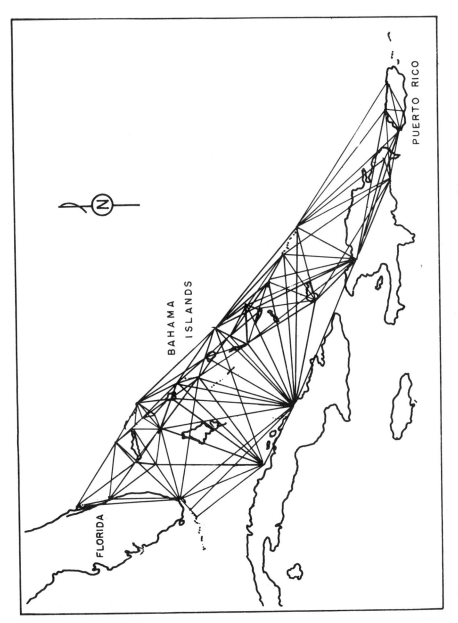

Figure 7–15. Shoran trilateration network extending from Florida through Puerto Rico.

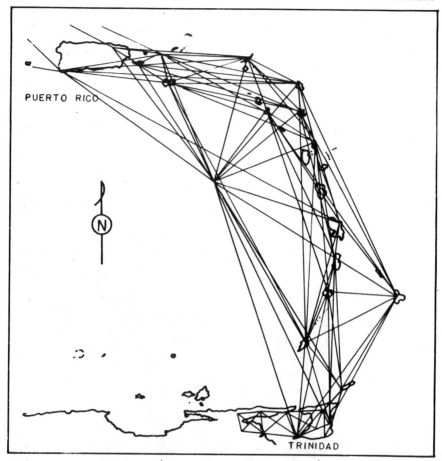

PUERTO RICO

TRINIDAD

Figure 7–16. Continuation of Shoran trilateration network.

eclipse projects had met with only limited success because of the fickle weather conditions encountered. Therefore, the Shoran system was brought into service. Ground stations were established across the Atlantic Ocean as shown by Fig. 7–17. The longest measured line was 549 miles and the shortest was 44 miles. The project was divided into five phases: the Norway–Scotland tie, the Scotland–Iceland tie, the Iceland–Greenland tie, the Greenland ice cap crossing, and the Greenland–Baffin Island tie. Work was begun in July 1953 and completed in August 1956. The precision of a station position in Norway with respect to a fixed position on Baffin Island is [20]:

probable error in latitude = ±0.0056 mile,
probable error in longitude = ±0.0049 mile,
probable error in the resultant = ±0.0074 mile, or about 39 feet.

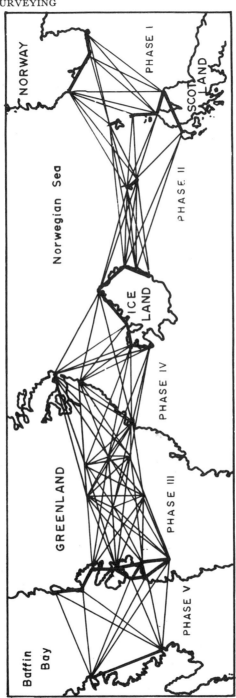

Figure 7–17. The North Atlantic tie.

The vast area of Canada lends itself superbly to Shoran trilateration [15]. A project was initiated in 1949 and completed in 1957, covering an area of two and one-half million square miles. The Shoran network consists of 143 stations with 502 measured interconnecting lines, of which the longest is 367 miles. Canada now has complete geodetic coverage in the area north of the present conventional triangulation system. The Arctic islands are integrated with the mainland and all results are on a uniform datum. The standard of accuracy in length is assessed as 1 part in 56,000, which has been confirmed by a limited number of Shoran stations which were subsequently incorporated into the conventional triangulation network for purposes of checking.

The list of Shoran accomplishments is long and these examples are typical. One famous project, the Marshall Islands survey, will be explained later in this chapter because it illustrates the case of using a number of different systems, each selected to perform that portion of the task to which it is best suited. Recently, Hiran was used to connect Australia to the Marshall Islands, with a side network which terminated in the Fiji Islands. This project, completed in 1965, illustrates the mobility of this system.

The errors cited above are typical for the system. The probable error of a single observed distance is ± 0.0024 mile, or about 13 feet. Twelve independent observations of each line are made and the entire network solved in a least squares adjustment, generally incorporating other independent observations such as precisely measured baselines and azimuth observations. In practice, the goal is to duplicate the design accuracy of the United States national survey system as estimated by Simmons of the Coast and Geodetic Survey; the formula is

$$\text{Proportional part accuracy} = \frac{1}{20,000\, D^{1/3}},$$

where D is the distance in miles along the axis of the net. This means that a network extending 1000 miles should have a probable error at the terminating position not to exceed 1 : 200,000, or $\pm$ 0.0050 mile, or 26 feet. For a survey of 2000 miles in length, the probable error of the terminal station should not be greater than 1 : 250,000, or ± 0.0080 mile, or 42 feet [12].

Gee-H. Another circular system is the Gee-H system designed by the British early in World War II primarily as an air navigation and bombing system. The name "Gee" comes from the first letter of the word "Grid." The H is supplied because the system measures the distances from an aircraft, or surface vessel, to two fixed ground stations. Such systems are known as H-systems, and include Gee-H, Shoran, and the Electronic Position Indicator. By contrast, the system that measures the distance from two ground stations to the mobile station is called "Oboe."

Gee-H was adapted to aerial surveying before the end of the war in the preparation of tactical maps in Southeast Asia. Techniques were perfected and the system was then used in photogrammetric mapping along the Gold Coast, Gambia, Sierra Leone, and Tanganyika [11].

The airborne station consists of a pulse transmitter and receiver. Like Shoran, pulses are sent alternately to each ground station at frequencies in the 80 to 100 MHz band. The ground stations respond on frequencies from 20–40 MHz. Unlike Shoran, which uses a circular time base, the Gee-H time base is linear. On the face of the CRT, two sets of time bases appear simultaneously, each time base corresponding to its respective ground station. Each time base consists of three different scales: the main scale which is 25 Gee-units in length, and two expanded scales. A Gee-unit is defined as the distance traversed by radio waves in 66.66 microseconds, or slightly less than 20,000 meters for the two-way loop distance, or about 10,000 meters for the measured distance. As in Shoran, a recording camera photographs the face of the CRT and is synchronized with the aerial survey camera so that the photographs are made simultaneously. Measurements are made later from the recorder films under magnification which furnishes the information necessary to determine the spatial position of the aircraft at the instant the aerial camera was triggered. Experience indicates that the total internal standard error of one Gee-H air to ground distance measurement is about ±45 meters [11].

Oboe. This system, like Gee-H, is of British origin and was designed as a bombing system, its name coming from *o*bserved *b*ombing *o*f *e*nemy.

Like Shoran and Gee-H, Oboe is a circular system but, unlike these two, it carries the responder beacon in the aircraft while the two measured distances are controlled from the two fixed ground stations. One ground station is called the tracking, or "cat" station, while the other is known as the releasing, or "mouse" station. The aircraft flies a course which is a circle with the cat station as a center. This circle is predetermined to pass through the area where the aerial survey is to be made. The exact points along this circle where exposures are to be made are defined by their distances from the mouse station. The cat and mouse stations of the Oboe system are identical in function to the drift and rate stations, respectively, of the Shoran system when circular patterns are flown by a Shoran-equipped aircraft. However, circular patterns are seldom used in Shoran surveying, the two methods of operation being straight flight lines by the use of the Straight Line Indicator in the photogrammetric mode, and the figure eight line-crossing pattern in the geodetic mode.

Because Oboe uses much higher frequencies than Gee-H, and also because the measuring equipment is on the ground where it can be made much heavier and more sophisticated, the measuring accuracy is about twice that of Gee-H. On the other hand, the measuring problem is complicated because part of the necessary information, such as the instant of aerial camera exposure and the corresponding aircraft altitude, is in the airborne unit. This requires a more complicated synchronization system. Of the circular systems, Gee-H has, for these reasons, been adopted by the British for practical survey work. For scientific purposes, such as the determination of the propagation velocity of electromagnetic waves at different altitudes, Oboe is preferred because of its higher accuracy [11].

Electronic Position Indicator. This system, popularly known in its abbreviated form EPI, was developed between 1944 and 1947 by the Radiosonic Laboratory of the U.S. Coast and Geodetic Survey for use in hydrographic surveying. Loran, which comes from *long range navigation*, covers vast areas at sea because it uses low frequency pulses which are not limited to line of sight but hug the surface of the earth. Loran, however, does not furnish the accuracy needed in hydrographic work. Other more accurate systems, including Shoran and Hiran, have the necessary accuracy but are limited to line-of-sight propagation. Even when the ground stations are placed upon the highest available terrain, the coverage at sea is limited.

To fill this gap, EPI was designed to combine the low-frequency pulse characteristics of Loran with the distance measuring techniques of Shoran. In 1945, field tests of a prototype proved that the development objectives were sound. By 1947, the first production model was completed and tested in the Gulf of Mexico. It was later used extensively in the Gulf, followed by combined Shoran–EPI operations in the Bering Sea where lines up to 500 miles long were measured with EPI.

The principles of operation of EPI are similar to those used in Shoran. Radio pulses are transmitted from the ship to two ground stations where they are received and retransmitted back to the ship. The elapsed time for the loop distance is measured in microseconds. The intersection of the two time circles establishes the position of the ship. Only one frequency of 1850 kilohertz (KHz) is used for all transmissions. This fact requires a special delay system to distinguish between pulses from the two ground stations.

Distances are measured by use of three phase shifters designed to accept timing frequencies of 100, 10, and 1 KHz to correspond to time scales of 10, 100, and 1000 microseconds, respectively. Timing is controlled throughout the system by a precision quartz crystal oscillator. The standard error of one distance measurement as proved by laboratory and field tests is ±0.27 microsecond, which means ±40 meters in a measured distance. In the Bering Sea operation, comparison of the EPI measurements with lines of known length showed an average correction of 53.7 meters to be applied to the measured lines for a standard error of 1 part in 7000.

While several of the hyperbolic systems, such as Decca and Raydist, can be used in the circular mode, their operational principles are considerably different from those of the four circular systems described above. They will be described in their more appropriate category of hyperbolic systems.

HYPERBOLIC SYSTEMS

A hyperbola is the locus of points the difference of whose distances from two points is a constant. The two fixed points are called the foci of the hyperbola.

In electronic surveying, radio stations are located at the fixed points, or foci, and are abitrarily designated stations A and B as shown by Fig. 7–18. A mobile station is introduced into the system and designated

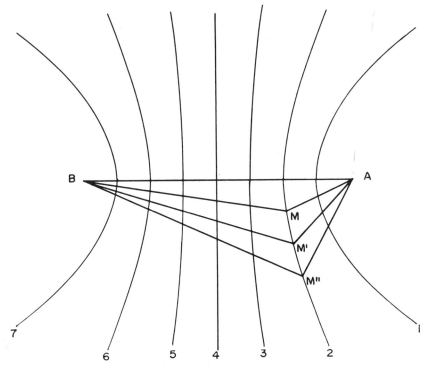

Figure 7–18. Pattern of hyperbolas with respect to baseline AB.

station M. If the mobile station M moves from point M to M' and thence to M″ in such a manner that

$$MA - MB = M'A - M'B = M''A - M''B = \text{a constant,}$$

it has, by definition, moved along a curve which is a hyperbola. The same procedure holds true for adjacent hyperbolas. The numbering and intervals of the adjacent hyperbolas are dependent upon the design of the specific electronic system under consideration. For simplicity in Fig. 7–18, the lines are numbered 1 through 7. Line 4, which is perpendicular to line AB and midway between the two stations, is a special case of the distance difference being equal to zero. With respect to the AB hyperbolic coordinate system, the coordinate value of the mobile station M is 2.00, or $M_{AB} = 2.00$. This tells us that the mobile station is somewhere along hyperbola 2 but does not provide a definite fix.

To get a definite location of station M, another pair of fixed stations must be established to furnish a second pattern of hyperbolas. In actual practice, only one extra station C is established because it can share station A as a mutual station with the first system.

In Fig. 7–19 a second pattern of hyperbolas is added with respect to stations A and C. In practice, these two patterns are differentiated by various designations, one favourite method being to call one pattern green, the other red. For convenience in this illustration, pattern AB has solid lines and pattern AC has dashed lines. With respect to pattern AC, mobile station M has a coordinate value of 8.00, or $M_{AC} = 8.00$. The coordinates of the mobile station M then become

$$M_{AB} = 2.00 \quad \text{and} \quad M_{AC} = 8.00.$$

There is a possible ambiguity in the location of M because the hyperbolas 2 and 8 intersect on the back side of the baseline BAC at point m. However, in practice these points are separated by sufficient distance to avoid confusion. Furthermore, for hydrographic operations, the stations are generally located along the coast in a geometric arrangement designed to provide specific coverage at sea on the concave side of the baseline BAC. Should a ship suddenly find itself at point m, the ex-skipper just established nautical history.

Hyperbolic systems may be either the pulse type or the phase-difference type. The pulse type is based on time differences, and systems belonging in this category are primarily designed to serve navigation and reconaissance rather than accurate mapping. Gee and Loran belong to this type. The phase-difference type is based on the principle that distance differences from a mobile station to two fixed stations can be determined by measuring phase angles of two waves transmitted from the fixed stations when the initial phase angles and the

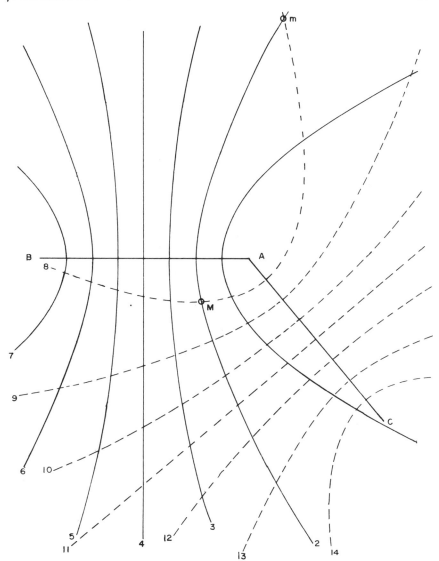

Figure 7–19. Hyperbolic location.

number of cycles included in the distance are known. The curves of constant phase difference are hyperbolas. These hyperbolas remain stationary relative to the two fixed stations as long as the propagation velocity, the transmitting frequency, and the initial phase relations at the fixed stations remain constant. Lorac, Raydist, and Decca belong to this type.

The fact that curves of constant phase angle difference are also curves of constant distance difference can be shown with the aid of Figs. 7–20 and 7–21. Radio wave length can be expressed in terms of the propagation velocity by the simple relationship

$$\lambda = \frac{V}{f}, \qquad (7\text{-}17)$$

where λ is the radio wave length in linear units, V is the velocity of electromagnetic waves, and f is the frequency of the transmission. This is shown graphically by the solid sine curve in Fig. 7–20 where the wave length λ is divided into units of phase angle expressed in radians.

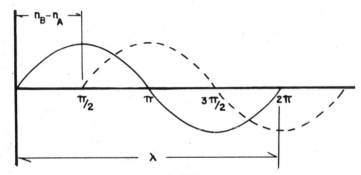

Figure 7–20. Wavelength and phase angle difference.

Phase angle is observable; however, from a single transmitter it will obviously vary in time with frequency. In essence, such a phenomenon from a single station produces little of geometric worth. On the other hand, consider the situation shown in Fig. 7–21, where the length of baseline AB is known. Station A transmits on a continuous frequency which is received at mobile station M. The same signal is also received at station B and retransmitted unaltered.

In terms of λ and parts thereof,

$$MA = \frac{n_A \cdot \lambda}{2\pi}, \qquad (7\text{-}18)$$

and

$$MB + AB = \frac{n_B \cdot \lambda}{2\pi}, \qquad (7\text{-}19)$$

where n represents the number of cycles. From these two equations,

$$MB - MA = \frac{(n_B - n_A)\,\lambda}{2\pi} - AB. \qquad (7\text{-}20)$$

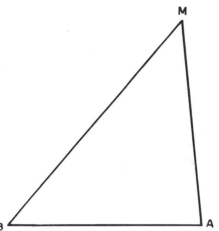

Figure 7–21. Single baseline with mobile station.

The measured phase angle difference, $n_B - n_A$, is shown in Fig. 7–20 and is a constant. As written, this observation represents part of a single cycle; whole cycles are determined by other means.

With a third station in the net, a second baseline is produced, as shown in Fig. 7–19. From this additional baseline, a second distance difference is similarly deduced and M is located at the intersection of the two hyperbolas thus defined.

Hyperbolic systems have been adapted to many different tasks throughout the years. Some of the more common uses are:

horizontal control for hydrographic surveys,
submarine cable laying,
horizontal control for dredging operations,
position fixing of navigation buoys,
speed and tactical trials of ships,
horizontal control for aerial photogrammetry,
position fixing for photo reconnaissance,
evaluation of missile guidance systems,
horizontal positioning of mobile missile firing platforms, and
horizontal control of airborne magnetic surveys.

Gee. This system was developed by the British in the early 1940's. The accuracy it provides is adequate for navigation and reconnaissance but is not sufficient to meet the needs of geodetic surveying or photogrammetric mapping and, therefore, will be treated briefly.

An arrangement of four stations, called the "star chain," is normally used. The master station is at the center of a rough circle with three stations placed around it along the circumference of the circle. In each star chain, the stations transmit on a common frequency between 20 and 85 MHz. Because of this common frequency within a star chain, a unique discrimination system using smaller "ghost" pulses is used for separation and identification of pulses from each of the three slave stations. Operating in this frequency range means that the distance between stations can be very little more than the optical line of sight. This puts the system at a disadvantage for accurate surveying because of the short baselines.

Loran. Development began in 1940 to establish synchronized pairs of high-frequency pulse-transmitting stations separated by several hundred kilometers to provide long range hyperbolic navigation to distances of 800 kilometers for high-flying aircraft. In the succeeding years, tests were made to find the band of frequencies which gives the best combination of ground-wave and sky-wave propagation by utilizing not only direct waves but also reflected waves from the ionosphere. The tests revealed that the lower frequencies produced the most stable sky-wave signals at night, whereas the higher frequencies gave more stability during the day. A compromise solution for both night and day long range navigation resulted in the present Loran medium frequencies between 1700 and 2000 KHz. This frequency range is now known as the Standard Loran Band.

Loran stations are generally installed as a chain of several stations along a coast line. The optimum distance between stations separated by sea water is about 500 kilometers; therefore, the entire Loran chain may extend over several thousand kilometers. This means that several independent pairs of stations must operate simultaneously without ambiguity. Four different channels within the Standard Loran Band are used. Each channel consists of eight pairs of stations which are identified by a special arrangement of pulse recurrence rates.

When using baselines 500 kilometers long, the average position error is about ± 2 kilometers at a distance of 1200 kilometers during the day. At night, the average position error at distances of 500 and 2500 kilometers are ± 2.5 kilometers and ± 14 kilometers, respectively [11]. While not accurate enough for geodetic or mapping purposes, Loran has firmly established itself as a universally used, long range electronic navigation system for both ships and aircraft.

Lorac. Lorac, an acronym for *lo*ng range *ac*curacy, is a phase-comparison radio-location system developed by Seismograph Service Corporation of Tulsa, Oklahoma. It is designed to provide continuous position information to distances of about 300 miles with accuracies from 10 to 200 feet. Operating at frequencies between 1600 and 2500 KHz, it uses only the ground-wave propagation and, consequently, the coverage is more limited than that of Loran. On the other hand, it provides accuracies suitable for geodetic and mapping purposes.

Tests have been conducted over lines of known length in the Louisiana and Florida coastal areas and over desert and mountainous terrain in Arizona. The test measurement data were collected at a radio frequency of 2258 KHz, under favorable meteorological conditions, and with a multiple number of line measurements. The conclusions reached from these tests, based on lines from 40 to 568 miles in length, indicate probable accuracies of from 1 : 25,000 to 1 : 100,000 for the coastal paths and from 1 : 10,000 to 1 : 75,000 for the desert and mountain paths.

When a particular line or group of lines needs to be occupied with a high degree of accuracy, such as the paths to be followed in submarine cable laying, hydrographic reconnaissance, or aerial photography, an auxiliary device known as Actrac is used. (Actrac is derived from *ac*curate *trac*king.) It is a plotting instrument driven by two Lorac indicators. One indicator drives the strip chart and the other indicator drives the pen. The trace made on the chart is a reproduction of the path made by the mobile station in the Lorac grid. Corrections to the path of the mobile station are made to cause the actual path to coincide with the desired path.

Of particular interest is the fact that Lorac systems were established in the vicinity of Cape Canaveral (now Cape Kennedy) and Patrick Air Force Base in 1957 for the purpose of positioning ships and submarines during the development of the Fleet Ballistic Missile known as Polaris. These mobile missile platforms stood offshore about 40 miles and lofted their warheads down the Atlantic Missile Range (now the Eastern Test Range). Lorac was also used on the range during submarine cable-laying operations and in the calibration of hydroacoustic missile impact location systems. Two Lorac systems continue to provide operational support for the Air Force and Navy.

The Pacific Missile Range installed a permanent Lorac system in 1964 in the vicinity of Point Mugu, California, and the Naval Undersea Warfare Center is currently installing a Lorac system in the vicinity of Long Beach, California. The Seismograph Service Corporation maintains complete coastal coverage of the Gulf of Mexico.

Resolution of position ambiguity within the Lorac system is provided by a recently developed two-stage lane identification system which eliminates the necessity for ships to return periodically to fixed reference points. Remote controlled operation of base transmitting stations is included as an integral part of the lane identification system. Representative Lorac equipment is shown in Figs. 7–22 and 7–23.

Other recent developments by Lorac are: (1) a Zero Header to provide a readout of the ship's distance in feet from a point which it desires to occupy. This equipment is also an effective assistant to the

Figure 7–22. Left, Lorac reference transmitter; right, Lorac base transmitter (Seismograph Service Corporation).

helmsman in maintaining a position after it is occupied. (2) a Digital Recording System, shown in Fig. 7–24, which transforms the Lorac analog output to digital form for recording on magnetic tape, teletypewriter, punched tape and digital printer.

Raydist. Raydist is a development of Hastings-Raydist, Inc., of Hampton, Virginia. The name is a contraction of *ray*-path *dist*ance. It is a continuous-wave heterodyne phase-comparison system of radio location designed for air, land, and sea navigation. Raydist was first employed successfully in 1940 at Langley Field, Virginia, as a system for precise measurement of the speed of aircraft. After World War II, it came into extensive use in off-shore oil exploration and drilling operations in the Gulf of Mexico. The Air Force and Navy have used Raydist in automatically tracking and positioning aircraft. The Lincoln Laboratory of the Massachusetts Institute of Technology has used it for tracking aircraft in connection with radar evaluation studies. For a number of years it has been used by the U.S. Army Corps of Engineers in dredging operations. Other applications have been ship-speed

Figure 7–23. Lorac shipboard installation (Seismograph Service Corporation).

testing, hydrographic mapping, charting of rivers, channels, and coast-lines, and navigation of ships through narrow channels and shoals.

Raydist determines distances or differences of distances in terms of phase shift. Continuous-wave transmitters emit radio signals which differ from each other by audio frequencies. These heterodyne audio frequencies are received at two or more points where, by comparing the phase of the signals as received, distance determinations are made.

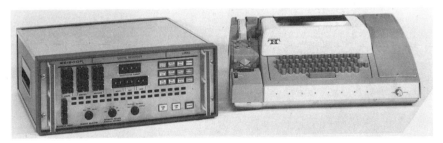

Figure 7–24. Digital data recording system with teletype printer (Seismograph Service Corporation).

Operating as it does in the 1800 KHz frequency band, Raydist is not limited to line of sight. A 10-watt system is used for close-in shore work up to 25 miles, while a 100-watt system is available for operating ranges up to several hundred miles [11]. Under carefully controlled conditions, accuracies equivalent to one part in 90,000 were obtained by the National Bureau of Standards.

This versatile electronic system was originally designed as a hyperbolic system. Other geometric patterns have been developed, making it possible to obtain precision through much broader areas with much less dilution of accuracy. These are the two-range pattern which makes Raydist similar to the circular systems previously described, plus the circular-hyperbolic and the hyperbolic–elliptical patterns which are described later in this chapter.

Newest in the Raydist family is the DR–S system. It is a single-sideband system which reduces the number of required frequencies to two of narrow band. It can be a multi-party user system and does not require transmission between the base stations, thereby allowing greater baseline lengths and increased operating ranges. Although basically a two-range system, it can also be converted to hyperbolic patterns. Figure 7–25 shows a Raydist DR–S base station. Figure 7–26 is a photograph of a Raydist navigator and automatic position plotter installed in a launch.

Figure 7–25. Raydist DR–S base station (Hastings–Raydist, Inc.).

Figure 7–26. Raydist navigator and automatic position plotter installed in a launch (Hastings–Raydist, Inc.).

Decca. Closely paralleling in time the development of Shoran was the development of Decca. Although the principles of the system were originally conceived in the United States in 1938, little commercial or military interest was shown in the invention. With the help of the Decca Record Company in London, a demonstration was held in 1940 before British officials in the United States. This was followed by trials for the British Admiralty in 1942, which led to the first operational use of the system by the Royal Navy as a mine-sweeping aid just prior to the D-Day landings. This success prompted the formation of the Decca Navigator Company in 1945. Permanent Decca navigation chains now cover the entire British coastline, a large portion of the western European seaboard, the Bombay and Calcutta areas in India, and the northeastern coast of Canada [11].

While Decca got its start as a navigation system, it has been identified increasingly with geodesy for such purposes as hydrographic surveying, gravity point measurements, marine oil exploration, aerial photography, and aerial magnetometer surveying. The first Decca chain to be operated for surveying purposes was established by the Royal Dutch Navy in 1947. This was followed by its adoption in other countries, so that it is now the standard tool in the hydrographic surveys of Great Britain, Japan, Canada, New Zealand, Denmark, South Africa, France, Holland, and Sweden.

Decca is basically a hyperbolic system for use on land, at sea, and in the air. The transmitting stations forming a Decca survey chain, comprising the master and the "red" and "green" slave stations, radiate nondirectional, unmodulated, continuous-wave signals in the 70 to 130 KHz frequency band. The three stations of a given chain are almost identical; the only important differences are the frequencies of the transmissions and the nature of the signal sources. The master transmitter is driven by a crystal oscillator of very high frequency stability and each slave transmitter is also driven by a local oscillator through a circuit which permits the transmitted slave signal to be locked rigidly in phase with the received master signal [5].

The Decca pattern thus produced is hyperbolic in form. Each family of hyperbolas is made up of lines of equal phase difference between the signals received from a pair of transmitting stations situated at the focal points, the continuous-wave transmissions being synchronized or "phase-locked" and assumed to travel at a constant speed. Given these assumptions, a comparison of the phase of the signals at the point of reception will locate the receiver on a hyperbola along which all points have a certain constant distance difference from the two stations. The indicated hyperbola forms a navigational line of position. Another pair of signals gives a second position-line by a similar process, intersecting the first at the receiver and fixing the user's position. The receiver drives two Decometers, each of which displays the phase difference between signals of a common frequency derived from the master and its associated slave transmission [5].

In the late 1950's, the two-range Decca system was developed. Whereas the conventional Decca system generates two families of hyperbolic position-lines, the two-range Decca system provides circular fixes as do Shoran and the other circular systems. The red and green slave stations are sited on the coast and the master is placed aboard the survey ship. The Decometer readings are then a function of the distance from the ship to the shore stations. The position of the ship is at the intersection of two circles centered on the two shore stations.

Signals transmitted at the working frequency of Decca have the property of closely following the curvature of the earth's surface with very little attenuation. The result is that surface vessels may use the system to great distances. Range is unlimited by the horizon and is independent of antenna height. The normal working range of Decca is about 150 miles, although satisfactory results have been obtained up to 1000 miles.

The standard deviation is specified as 0.01 of a mean lane. The change in distance along a baseline corresponding to a change of 0.01 mean lane is approximately 5 meters. With reasonable station geometry, it may be expected that errors will not exceed 60 meters anywhere in the survey area. A Decca Hi-Fix transmitter installation is shown in Fig. 7–27. A shipborne receiving system with continuous visual display is shown in Fig. 7–28.

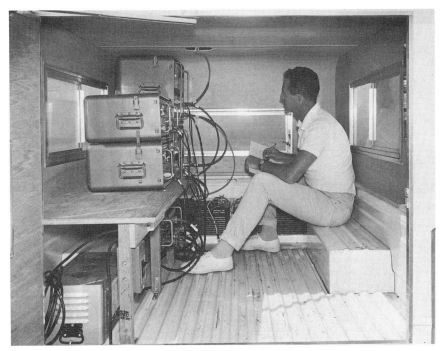

Figure 7–27. Decca Hi-Fix transmitter installation (Decca Survey Systems, Inc.).

CIRCULAR-HYPERBOLIC SYSTEMS

Several baseline lengths away from the baseline, the standard hyperbolic pattern degenerates rapidly in accuracy. This can be observed by reference to Fig. 7–29. The areas of uncertainty change from small squares or rectangles near the baseline into diamond-shaped areas of increasing size outward from the baseline. The accuracy of a fix in any of these areas may be evaluated as roughly proportional to the long diagonal of the geometric figure bounded by the lines of position generated by the system.

To overcome this deterioration in accuracy with increasing range, the circular-hyperbolic pattern was devised. The lattice, shown in Fig. 7–30,

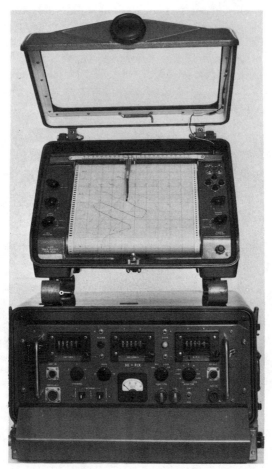

Figure 7–28. Decca shipborne receiving system with continuous visual display (Decca Survey Systems, Inc.).

is formed by producing a family of hyperbolas from one baseline of two stations, plus a family of concentric circles, or range-only measurements, centered on one of the stations. The resulting areas of uncertainty are rectangles of increasing length, but of fairly uniform width outward from the baseline. This pattern permits much more accurate coverage over broader areas than does the standard two-baseline hyperbolic pattern.

Systems which have been adapted to the circular-hyperbolic pattern are Raydist and Decca. Lorac has also been used in conjunction with Shoran to produce a circular-hyperbolic lattice. One occasion was during the calibration of the hydro-acoustic missile impact location

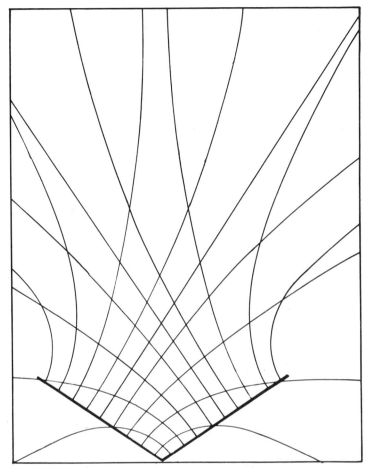

Figure 7–29. Standard hyperbolic pattern showing degeneration in accuracy away from the baseline.

system off the west coast of Ascension Island in 1957–1958. Two Lorac stations were established on Ascension Island to provide one family of hyperbolas. The available real estate did not allow a suitable location for an additional station to provide the customary second family of hyperbolas. Therefore, one Shoran ground station was established upon the top of Green Mountain to furnish the range to the calibration ship which contained both a mobile Lorac station and a mobile Shoran station. The ship was positioned laterally by reference to the single Lorac pattern of hyperbolas and in range by reference to the measured Shoran distance. In this instance, the marriage of two different systems produced the desired results.

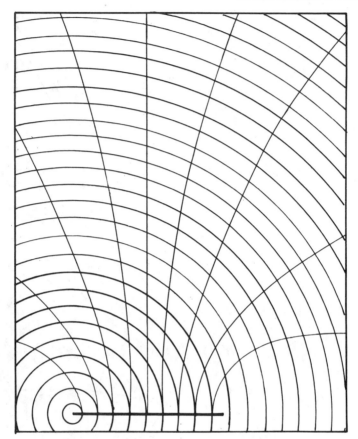

Figure 7–30. Circular-hyperbolic pattern.

HYPERBOLIC-ELLIPTICAL SYSTEMS

The rapid deterioration of accuracy with increasing range in the standard hyperbolic pattern has led to the development of another lattice known as the hyperbolic–elliptical pattern. Figure 7–31 shows this pattern. One family of hyperbolas is generated from one baseline of two stations by distance differences. However, there is no second baseline as in the case of the standard hyperbolic systems. Instead, using the same baseline which generated the family of hyperbolas, a family of ellipses is produced by using the constant distance sums. Outward from the baseline, the areas of uncertainty take the shape of rectangles of fairly uniform width and increasing length. As in the circular-hyperbolic case, the hyperbolic-elliptical pattern furnishes better accuracy over a broader area than is possible from the conventional hyperbolic pattern.

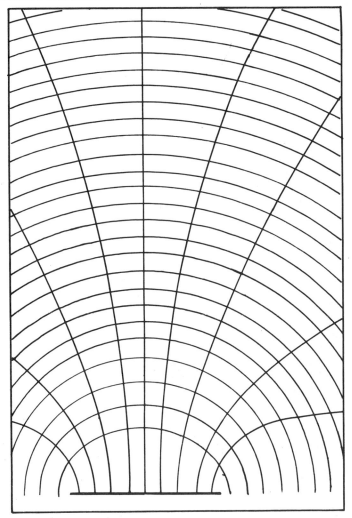

Figure 7–31. Hyperbolic-elliptical pattern.

At least one system, Raydist, uses the hyperbolic-elliptical pattern in one of their several optional modes of operation.

BASELINE MEASURING SYSTEMS

Some systems are specifically designed for the precise measurement between two points on the earth's surface and for this reason are used primarily for measuring baselines. However, their use has been broadened to include geodetic traverses, and also trilateration in networks of extreme precision such as the Southeastern United States Survey

(SEUSS). Included in the category of baseline measuring systems are the Electrotape, the Geodimeter, the Tellurometer, and the Väisälä light interference comparator. The Geodimeter is more properly classified as an electro-optical system. The Väisälä comparator is not an electronic system, being purely optical in nature, but functionally it falls in the category of baseline measuring instruments and will be discussed briefly.

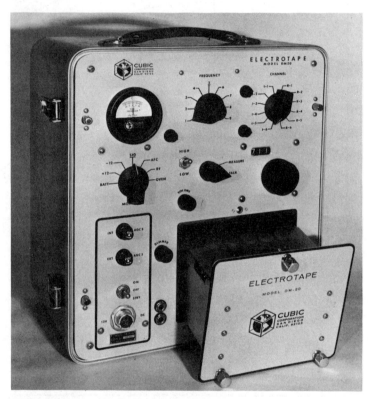

Figure 7–32. The Electrotape (Cubic Corporation).

Electrotape. The Electrotape, shown in Fig. 7–32, is a product of the Cubic Corporation of San Diego, California. It is a precision electronic distance-measuring instrument used to measure the slope distance between two points. Although distances greater than 30 miles have been measured successfully, the recommended distances to be measured are between 30 feet and 30 miles. The speed and accuracy of measuring with the Electrotape are outstanding even in rough terrain.

The Electrotape system consists of two distance-measuring units operating together as a pair. One unit is placed over the measurement point at each end of the line to be measured. Each unit transmits microwave signals to the other to establish contact. Under control of the operator, each unit can be operated either as an interrogator or a responder. During a typical operation, the first measurement is made with one unit operating as the interrogator and the other as the responder. A total of 28 counter readings are made. Then the function of each unit is reversed and a second measurement is performed, also with 28 counter readings. These two measurements comprise a set from which the slope distance is obtained. For an immediate on-site accuracy check and for coordination of field activities, Electrotape's built-in two-way FM radio provides direct contact between operators. The entire two-way measurement, including set-up time, is completed in about 15 minutes. For maximum speed and efficiency, the units are employed in a "leap-frog" manner.

Each unit is contained in a rugged, epoxy-fiberglass case. In operation, each unit is mounted on a tripod. A psychrometer and pressure–altitude barometer are furnished for use in precise distance measurements. The carrier frequency is from 10.1 to 10.4 gigahertz (GHz). The specified accuracy is 1 centimeter ± 1 part in 300,000, which has been confirmed by field testing.

Use in the field has demonstrated that Electrotape operation is independent of daylight, rain, fog, snow, or other weather or climatic conditions. Temperature extremes do not affect performance. Precise measurements have been made in temperatures from 125°F in the desert to 40°F below zero in both the Arctic and Antarctic. The electronic circuitry is all transistorized; there are no vacuum tubes except the klystron. The precision crystals are temperature-stabilized in 185°F ovens.

Electrotape is now being used by numerous land surveying, civil engineering, aerial surveying, and photogrammetric firms, as well as by city, county, state, national, and commercial survey agencies. The system is ideally suited to traversing, trilateration, and polar surveying. In polar surveying, one triangulation station is occupied by a 1-second theodolite, a second station is sighted upon to provide the azimuth, then the angle is turned to the new horizontal control point. The distance from the first station to the new control point is then measured with the Electrotape. This procedure is continued for the establishment of additional control points. For distances up to about five miles, the polar method is generally faster and less susceptible to errors than either traversing or trilateration. The disadvantage is that good visibility is required for turning the angles [22].

In particular, the Electrotape is being used to great advantage in areas of California adjacent to the Pacific Ocean. Coastal fog and the

Figure 7–34. Rear view of the Model 6 Geodimeter (AGA Corporation).

Figure 7–33. Front view of the Model 6 Geodimeter (AGA Corporation).

resultant haze and smog are often too severe to obtain the accuracy required in angular measurements. Often conventional theodolite or transit survey parties arrive on station only to find the visibility so poor that the resulting measured angles are unacceptable. As a relief from this problem, many survey agencies have turned to the Electrotape with excellent results, from the standpoint of both accuracy and speed of operations [13], [21], [22].

Geodimeter. The Geodimeter, one model of which is shown in Figs. 7–33 and 7–34, is the invention of Dr. Erik Bergstrand, geodesist of the Swedish Geographical Survey Office. In 1941 he began experiments to determine a better value of the velocity of light so that he could develop an accurate distance-measuring device. From his research came the Geodimeter, a word derived from *geo*detic *di*stance *meter.*

The method used by Bergstrand has its origin in the work of Fizeau and Karolus. In 1849, Fizeau made the first terrestrial measurements of the speed of light. His apparatus consisted of a light beam, a toothed wheel, and a distant mirror. The light beam was directed through the cogs of the wheel to the distant mirror where, after reflection, it was observed again through the cogs. When the wheel reached a certain angular velocity, the reflected light beam was eclipsed so that the observer could not see it. The velocity of light could be computed by knowing the distance to the mirror, the angular velocity of the wheel, and the dimensions of the cogs [11].

Proceeding from this arrangement of Fizeau, Karolus and his co-workers carried out a series of experiments between 1925 and 1940 in which the rotating cogwheel was replaced by a noninertial method of light modulation. By means of the electro-optical Kerr cell, a light beam was modulated up to a frequency of 10 MHz, thus creating an extremely fast electronic shutter [10]. This modulation of the steady output of a light source produces a beam whose intensity varies as a sine curve of known wavelength. This so-called "blinking" light is projected to a distant reflector of which there are three types: plane mirror, spherical mirror, and tetrahedron prisms mounted in a housing. Figure 7–35 shows the latter type. The light returned from the reflector is received by a photo-tube in the Geodimeter. Because the light takes a finite time to make its loop, the photo-tube receives light of a different intensity, or a different point on the sine curve representing intensity. This displacement in phase represents distance. The photo-tube converts light intensity into electrical current which is measured and converted to distance [7].

The Geodimeter is produced in several models, varying from heavy equipment for precision baseline measurements to lighter equipment designed for traversing, picture point surveys, and highway surveys.

Figure 7–35. Retro-directive tetrahedron prisms (U.S. Coast and Geodetic Survey).

Distance measurements from 3 to 30 miles or more are possible depending upon the model. Accuracies also vary according to the model. As an example, the mean error of a single observation with the Model 2A Geodimeter is given as 1 centimeter ± one-millionth of the distance, which gives an error ratio as shown in Table 7–3. Szabó [18] reports an accuracy of 1:800,000 for the Geodimeter when calibrated against accurate standard baselines.

The Geodimeter has enjoyed great popularity in this country. It has proved its accuracy, dependability, and ruggedness. Maintenance costs

TABLE 7–3

Error Ratio of the Model 2A Geodimeter versus Distance

Distance (km)	Error Ratio
1	1 : 100,000
5	1 : 330,000
10	1 : 500,000
20	1 : 700,000
30	1 : 750,000
50	1 : 830,000

have been low. Its principal disadvantage is that it is limited to night operations. Rain does not hamper measurements but fog puts a stop to operations.

One of the most famous surveys making use of the Geodimeter is the Southeastern United States Survey (SEUSS). This is the designation for the horizontal control centered around that portion of the transcontinental traverse extending from Aberdeen, Maryland, south to Cape Kennedy, Florida, and from a connection near Jacksonville, Florida, west and north to Greenville, Mississippi. It was along or near this survey that several geodetic satellite observing systems were placed and used for accuracy tests. Naturally, in testing this new equipment, there was a need for the best possible basic control for checking scale and orientation. Mr. Lansing Simmons of the U.S. Coast and Geodetic Survey believes that nominal accuracy for the Geodimeter traverse is one part in a million.

Tellurometer. The Tellurometer, derived from the Greek words for "earth" and "measurement," is the name applied to a system developed in South Africa for use in the South African Trigonometrical Survey. It has three forms: the original Microdistancer equipment for overland measurements, introduced in 1957; the Hydrodist equipment, just recently introduced and designed to the specifications of the South African Navy Hydrographer, to obtain the position of hydrographic survey vessels; and Aerodist, now under development for airborne applications. Aerodist will be discussed later in this chapter under New Developments.

The Tellurometer, in the Microdistancer form, was designed with three basic objectives in mind: accuracy, portability, and all-weather operations. In order to get around the disadvantage of using modulated light and consequently being limited to night operations, as in the Geodimeter, the Tellurometer uses radio frequencies of 3000 MHz (10-centimeter wavelength). This is modulated by a basic pattern frequency at 10 MHz and by three other frequencies close to the basic frequency. The system is both a phase- and pulse-measuring device. Pulses are initiated at the master unit, received at the remote site, analyzed, and retransmitted to the master unit. At the master unit, the pulses are presented on a small cathode ray tube for reading and recording. The indication on the master unit requires no manual operation such as matching pulses as is done with Shoran. Instead, a trace is read directly on a calibrated graticule fastened to the face of the cathode ray tube. The system measures the time required for an energy pulse to travel from the master to the remote unit and return. The time is measured in terms of a phase shift of the modulated wave; the unit of measurement is the millimicrosecond, or a billionth of a second. Once this time is measured, it is multiplied by the velocity of propagation and divided by two to obtain the length of the line.

Since propagation velocity varies with meteorological conditions, an index of refraction correction must be made. For these meteorological measurements, a barometer and sling psychrometer are provided. Voice communications are an essential part of field operations.

The maximum range of the Tellurometer is about 70 kilometers. Accuracy, as specified by the manufacturer, is 3 parts per million of the distance ±2 inches. The U.S. Coast and Geodetic Survey made a test over a line 12,800 meters long with a difference between the established length and the Tellurometer length of 6 centimeters, thus providing an accuracy of 1 part in 200,000. A second comparison was made over a line 2600 meters long and provided an accuracy of 1 part in 100,000. These two tests were made in the Washington, D.C., area under average field and weather conditions [6]. Other investigators report accuracies from 1 part in 61,000 to 1 part in 450,000 [4], [11], [18].

The Tellurometer, one model of which is shown in Fig. 7–36, has been used in nearly every part of the world. It has found application not only in baseline measurements but also in traversing and trilateration. One country making full use of the Tellurometer is Australia. In 1955, the managers of the Weapons Research Establishment in Australia very carefully weighed the merits of all electronic surveying systems for possible use in extending the geodetic control of their test range. The choice was the Tellurometer. Concurrently, the Commonwealth and State Survey authorities adopted the Tellurometer and put 40 units into service. By 1960, the total length of lines in Australia measured to an accuracy of 1 part in 100,000 was 900 miles [14].

The nautical version of the Tellurometer, called the Hydrodist system, is used for accurately positioning a survey ship. It comprises two single line-measuring instruments. The two master stations are located in the vessel, while the two remote stations are located at known positions ashore in such a manner as to form a triangle. Each master works continuously its own remote station. The vessel must have an uninterrupted line of sight to both land stations. Satisfactory operation up to distances of 20 to 25 miles are common with accuracies of from 1 part in 30,000 to 1 part in 100,000 [17].

Väisälä Light Interference Comparator. Although not an electronic device, no discussion of baseline measurements would be complete without including the light interference method designed in 1923 by Y. Väisälä and put into practice by the Finnish Geodetic Institute. By 1929, the instrument was fully developed and measurements of up to 200 meters were made with an accuracy of 1 part in 10,000,000 [9]. By 1947, the Nummela baseline in Finland was measured by Dr. Kukkamäki and

Figure 7–36. Side and front view of the Model 2 Tellurometer (U.S. Coast and Geodetic Survey).

Dr. Honkasalo with an rms error of 1 part in 17,000,000. This was followed by the measurement of a baseline of 480 meters in Argentina to an accuracy of 1 part in 9,000,000. Later, a 576-meter baseline was measured by the same two scientists in the Netherlands with an rms error of 1 part in 11,000,000. Dr. Honkasalo also participated in the measurements of the southern part of the Munich standard baseline in Germany where the accuracy was in the order of 1 part in 20,000,000 [8].

In principle, white light from a light source is passed through a collimator to make the rays parallel. The beam is divided at a mirror which has two holes in it. The first beam goes through a hole in the mirror to a distant mirror, where it is reflected back through the second hole and onto the focal plane of an observation telescope. The second beam is reflected between the mirror with the two holes and a solid mirror. The effective angle between these latter two mirrors is adjustable to very high precision by a compensator consisting of two plane parallel glass plates. By turning the angle of one of the plates, the second beam of light can be made, after a certain number of reflections, to appear in the focal plane of the observation telescope. The angle of rotation of the compensator is a measure of the number of reflections of the second beam and, hence, its distance of travel. The difference in the distance travelled by the two beams can be observed to an accuracy of one light wave on the basis of the symmetry of the diffraction interference lines as observed through the telescope.

Accurate measurement by light interference is a laborious task. Ground stability is extremely critical. Even bedrock is not stable enough for the baseline. In Finland, there is moraine gravel under the standard baseline to a depth of 25 meters. It has proved to be extremely stable through four independent measurements [9].

While not a tool the average geodesist will ever encounter, the light interference comparator has solved the problem of calibrating other baseline measuring systems. The Väisälä light interference comparator was adopted internationally in 1954 for uniform calibration of other baseline measuring instruments [18].

ALTITUDE MEASURING SYSTEMS

In electronic surveys on land where vertical control is available or can be established, or in hydrographic surveys where sea level is always available, the problem of determining the altitude of an electronic surveying system is simple. In the case of airborne electronic surveying systems, however, the altitude is much more difficult to obtain. The simple barometric pressure altimeter contained in the aircraft system has been used for some less precise surveying operations but it is not sufficiently accurate for geodetic purposes. Two instruments have been used almost exclusively in the air leveling method of determining vertical control: the frequency modulated (FM) altimeter and the pulse-type altimeter. These have come into general use in establishing vertical profiles in topographic surveying, supplying the vertical control for gravity and magnetometer measurements in areas where good basic control does not exist, and controlling long strips in aerial trilateration.

The Frequency Modulated Altimeter. The FM altimeter determines the height of the aircraft above the terrain by measuring the time which it takes for transmitted electromagnetic energy to reach the ground, be reflected, and be received back at the aircraft. The altimeter transmits a signal whose frequency is varied rapidly at a constant rate so that the signal which is received back at the aircraft has a different frequency from the signal then being transmitted. This frequency difference is a function of the length of time it took the signal to make the round trip.

The same familiar basic equation which is used in other electronic distance-measuring equipment pertains also to this system:

$$H = \frac{V \cdot T}{2}, \tag{7-21}$$

where H is the altitude of the aircraft above the terrain, V is the velocity of propagation of electromagnetic energy, and T is the time for the signal to make the round trip as determined from the frequency difference.

The FM altimeter is used mostly in low-altitude, slow-moving aircraft. The transmitter frequency is varied constantly in a sinusoidal manner between 420 and 460 MHz at the rate of 120 times per second. The output of the transmitter is 1/10 watt and is radiated from a dipole antenna located on the bottom of the aircraft. A separate receiving antenna is also located underneath the aircraft.

Depending upon the scale selected, the maximum error is from about 1.5 meters to 15 meters, $\pm 5\%$ of the altitude [11].

Pulse Type Altimeter. In this type of radio altimeter, bursts of electromagnetic energy are transmitted from the aircraft by a directional antenna located in a fixed position on the under surface of the aircraft.

Because of the highly directional nature of the antenna, the measured altitude must be corrected by the tilt of the aircraft in order to find the true altitude,

$$H = H_1 \cos t, \tag{7-22}$$

where H_1 is the measured altitude of the aircraft above the terrain, H is the true altitude of the aircraft above the terrain, and t is the angle of tilt of the aircraft.

The pulse-type antenna is in general use in high-altitude, fast-moving aircraft. Because of the high speeds, it has become more difficult to correlate the aircraft location with its altitude. Programs have been underway for some time to develop a system which will record a continuous profile of the actual flight path. Electronic Associates, Ltd., of Canada, has successfully built a system for the Royal Canadian Air Force known as the Airborne Profile Recorder (APR). It has found use not only in Canada but also in the United States. The Aerospace Cartographic and Geodetic Service (formerly the 1370th Photo-Mapping Wing), which has the primary photo-mapping mission for the Department of Defense, has equipped their RC–130 and RC–135 aircraft with the APR.

The APR transmits pulses of one-half microsecond duration at a wavelength of 1.25 centimeters and a pulse repetition frequency of 1000 cycles per second. The parabolic reflecting antenna, which is mounted underneath the aircraft, produces a conical beam with an effective vertex angle of about 1.2 degrees. The same antenna is used for both transmitting and receiving. The sweep frequency of the time base is 245.85 KHz and is generated by a crystal-controlled oscillator. The period produced by this oscillation is 4.068 microseconds, which is the time required for electromagnetic energy to make the round trip to the ground from a flying altitude of 2000 feet. This oscillator output furnishes extremely accurate increments of 2000 feet for the succeeding stages of the timing circuit. Altitudes up to 48,000 feet can be determined and presented on a graphic recorder.

The APR not only measures elevation but also corrects the flying altitude in relation to the isobaric surface. By linking a sensitive aneroid capsule to a parallel plate condenser, a change in capacitance is used to generate a direct current which is proportional to the pressure change. The output of this circuit is connected to a recording milli-ammeter so that it corrects the radar readings automatically to the selected equal-pressure surface. It can compensate for changes of altitude of ± 200 feet from the predetermined flight altitude [11].

Laurila [11] describes two tests. The first one was made in Canada. By comparing the APR elevations with known elevations, a standard error of one elevation determination was ± 3.7 meters. The second test was performed three times over mountainous country in Alaska. The standard error of the means compared with the true known elevation was ± 3.0 meters.

COMBINED SYSTEMS

In modern geodetic surveying operations, two or more systems are often used together to produce better results than could be obtained by any one system used alone. In a previous section devoted to circular-hyperbolic systems, reference was made to the fact that Lorac and Shoran have been used together, Lorac to produce a hyperbolic lattice from one baseline and Shoran to provide range from a single station. As another example, Shoran and a radio altimeter are normally used together. Shoran furnishes slant ranges from the aircraft to two points of known location, while the radio altimeter gives the altitude of the aircraft.

Often these modern surveys involve different categories of systems, such as optical, electronic, and gravimetric. A typical case is the famous Marshall Islands survey wherein 9 stations were tied together by a trilateration network of 32 lines, each measured at least 12 times by

Shoran. All 9 stations were located astronomically by optical systems. A baseline approximately 75 miles long between Ebadon and Kwajalein was measured by use of the Tellurometer. The entire network was oriented in azimuth by tracking a light on the Shoran aircraft with a theodolite at each end of a baseline as the aircraft performed line crossings. Gravity observations were made from a submarine in a systematic pattern around Wake Island and Eniwetok Atoll for the purpose of determining the deflection of the vertical.

The survey, shown by Fig. 7–37, was completed in 1960 and is one of the most sophisticated surveys ever made. It was designed with the intention of taking maximum advantage of the unique capabilities of each system. The excellent result obtained is indicative of what can be accomplished by intelligent use of combined systems.

RECENT PROGRESS AND NEW DEVELOPMENTS

Because of the relative newness of electronic distance measuring applied to geodetic surveying, it can be expected that many significant improvements will evolve each year. Research and development are continuing to produce new systems and to incorporate new techniques into present instruments. Several of these new developments will be explained.

In the United States, the Cubic Corporation has recently introduced a microwave system for hydrographic, oceanographic, geophysical and topographic surveys. Its name, Autotape, is derived from the automatic manner in which the system fixes positions of moving ships or helicopters with high-order accuracies. It is a transistorized, compact and portable system with an operating frequency of between 2900 and 3100 MHz. The system automatically measures the slant distance between two or more responders and visually displays those ranges in five metric digits on the instrument panel. The display may also be recorded in an unattended operational mode. Autotape uses the method of comparing the phase of modulated frequencies in sequence. Surface distances up to 30 miles can be measured to an accuracy of 50 centimeters plus 1 part in 100,000 of the range. By line crossings, operating distances can be increased. In a recent helicopter line crossing, measurements made at 124 miles were within 1.42 meters of the true distance. Real efficiency is combined with real portability. The largest of the Autotape units, the two-range interrogator, weighs only 55 pounds. The two types of antenna are extremely compact. The horn type weighs only 20 pounds while the omnidirectional antenna weighs 4 pounds and is only 18 inches long. Figure 7–38 shows a complete system consisting of the interrogator, one responder (for two-range use a second responder is

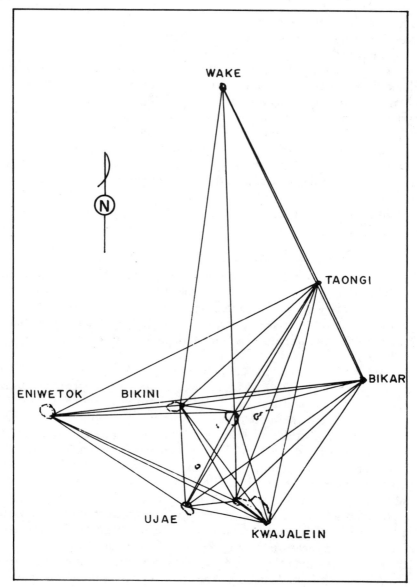

Figure 7–37. The Marshall Islands survey.

needed), a printer for automatic recording, and both the horn antenna and the omnidirectional antenna.

Another form of the Tellurometer system, called Aerodist, has been under development. Aerodist is a system used in line measurement by

Figure 7–38. The Autotape system (Cubic Corporation).

aircraft. Measurements to an aircraft 100 miles or more away have been made successfully. By combining two such measurements, lines over 200 miles in length can be measured. It can also be used as an aid in photographic survey. The overall accuracy of the Aerodist system is 1 meter ±1 part in 100,000 of the measured distance [17].

In 1966, a standard Model 4 Geodimeter was modified. Using a He–Ne Laser, a modulator, a red-sensitive photomultiplier, and modified optical coatings, an increase in sensitivity by a factor of two was obtained with the corresponding ability to measure under more difficult seeing conditions. The measurements made by this modified system were compared with those made by the standard Model 2A and Model 4 instruments. On lines 15 kilometers long, the results agree within better than 1 part in one million. The modified Model 4 has been used successfully on a 42-kilometer line [19].

Another recent development of the Cubic Corporation is Shiran, an acronym for S-band Hiran. The first part of the acronym is correct as the system does operate in the S-band; the remainder of the acronym is meaningless because Shiran is truly a new distance measuring system, but because Hiran, like its predecessor Shoran, has done so much for geodesy, it was decided to retain the word Hiran and prefix it with the letter S. Shiran is not at all like Hiran in appearance or principle. Development began in early 1961 because the extensive world-wide surveying programs planned for the future require a more efficient and accurate electronic surveying system. Requirements for Shiran can be summarized simply as (1) the capability of measuring the length of a line on the earth's surface having a maximum length up to 900 nautical miles (each distance measurement being approximately 450 nautical miles) with a probable error of ± 10.3 feet or less, and (2) the capability of determining the ground nadir of aerial photographs with ± 24 feet for 90% of the measurements. These two requirements are related to two principal types of missions used in airborne surveying, line crossings and photomapping [16]. In early 1968 the system was in the test stage with four units installed in RC–135A aircraft of the 1370th Photo-Mapping Wing. Testing has been accomplished over a six-station network extending from the southern part of Arizona to Central Nevada, plus one over-water line off Southern California. Indications are that the system is meeting specifications. This does not imply that 450 nautical mile distances are normally measured. Because propagation is line of sight, the range is determined by ground station angular view to the horizon, the aircraft altitude, and atmospheric conditions. In broad terms, Shiran is a circular system. It measures and records nautical miles from the aircraft to as many as four Shiran ground stations. The basic system consists of an airborne interrogator set, shown in Fig. 7–39, which transmits a continuous-wave, phase-modulated signal to each of four ground transponders, shown in Fig. 7–40, interrogating each transponder sequentially 10 times per second. Each transponder re-transmits the signal back to the airborne station which then automatic-ally measures the phase delay of the returned modulation signals and interprets the phase difference as range. All four ranges, plus time and mission identification data, are recorded on magnetic tape in a format suitable for computer reduction after each flight.

Investigations have been made in Sweden for new ways of making electro-optical distance measurements. From this work have come several new developments, one for Kerr cell operated light modulators, another for concentrated arc lamps, and still another for quartz-crystal modulation. From these developments, the Terrameter has been designed. It includes two main parts, the transmitter and receiver portion, and a mirror for reflecting the light signal back to the instru-ment. The transmitter emits light which, after polarization, is collected in a lens and concentrated in the center of the crystal. The outgoing light

Figure 7-39. Front view of the Shiran distance measuring interrogator set (Cubic Corporation).

from the crystal is then collected in another lens and made parallel. In the front of this lens there is a polarizer with two polarization axes perpendicular to each other. When a measurement is made, the last polarizer is operated so that readings are made by using both polarization planes. Maximum distances of about 3 to 5 kilometers can be measured. A test was conducted with 10 measurements over 1 kilometer. The standard deviation of a single measurement was ±11 millimeters [3].

Laser technology is only in its infancy but already lasers have been put to practical use. Spectra-Physics, Inc., of Mountain View, California, has developed a gas laser-transit for use in aligning dredging vessels working on San Francisco's new Trans-Bay Tube. The laser-transit consists of an ordinary surveying transit fitted with a gas laser mounted parallel to the telescope. A thin fan-shaped beam spreads 7 degrees vertically and 12 seconds horizontally. The beam can be rotated 90 degrees to provide either a vertical or horizontal reference line. The laser beams are projected over the water and the crews align

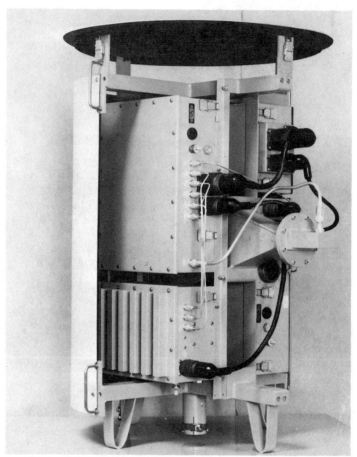

Figure 7–40. A Shiran transponder (Cubic Corporation).

the dredge visually with the beams. The beams are visible night and day, and are of low enough power to be eyeballed. Of course, fog is a limiting factor. The U.S. Coast and Geodetic Survey is conducting a series of laser tests using the Geodimeter as the basis of comparison. While no written report is yet available, a verbal report in early 1968 indicated that the interim results are very favorable. The Army Map Service is now conducting a study of aerial mapping by lasers. The concentrated light source accurately identifies ground elevations in dense tree growths. Their opinion is that a Laser Terrain Profile Recorder may soon be a practical instrument of the military map maker. On the basis of this optimism, it appears that lasers have a definite place in the future of surveying.

These are by no means the only developments which have been completed recently or are underway. They serve, however, to illustrate the general trend in equipment development.

Paralleling these efforts are studies of limitations from atmospheric turbulence, and the investigation of possible methods of compensating for propagation effects. These studies and investigations are vitally necessary if the ultimate in accuracy is to be derived from electronic surveying.

REFERENCES

1. Carl I. Aslakson, Some aspects of electronic surveying, *Proc. Am. Soc. Civil Engineers.* **77**, Separate No. 52 (1951).
2. Carl I. Aslakson, The velocity of light, *International Hydrographic Rev.* **XLI**, No. 1 (1964).
3. Arne Bjerhammar, An electro-optical device for measuring distance, *J. Geophys. Res.* **65**, No. 2 (1960).
4. G. Coets, Remarks concerning current use of the Tellurometer, *J. Geophys. Res.* **65**, No. 2 (1960).
5. Decca Navigator System Limited, The Decca Navigator System as an aid to survey, Issue 5, London, England.
6. H. P. Demuth, Tellurometer traverse surveys, *Tech. Bull. No. 2*, U. S. Coast and Geodetic Survey, March 1958.
7. *Engineering News-Record*, Distance measurement device uses light waves, reprint from issue of January 2, 1958.
8. W. A. Heiskanen, Geodetic base lines, *J. Geophys. Res.* **65**, No. 2 (1960)
9. T. Honkasalo, Measurement of standard base line with the Väisälä light-interference comparator, *J. Geophys. Res.* **65**, No. 2 (1960).
10. A. Karolus, Physical principles of the electro-optical determination of distances, *J. Geophys. Res.* **65**, No. 2 (1960).
11. Simo Laurila, *Electronic surveying and mapping*. Publication of the Institute of Geodesy, Photogrammetry and Cartography, The Ohio State University, Columbus, Ohio, 1960.
12. Samuel D. Owen, Evaluation of Hiran networks, *J. Geophys. Res.* **65**, No. 2 (1960).
13. Frederick G. Peters, Electrotape in trilateration, paper presented at the 1964 Regional Convention of the American Congress on Surveying and Mapping, Kansas City, Mo., September 24–26, 1964.
14. G. R. L. Rimington, Report on electronic distance measurements in Australia, *J. Geophys. Res.* **65**, No. 2 (1960).
15. J. E. R. Ross, Canadian shoran project, *J. Geophys. Res.* **65**, No. 2 (1960).
16. E. M. Salkeld, Development of a precise geodetic survey system, *IEEE Trans. Aerospace and Electronic Systems.* **AES–2**, No. 1 (1966).
17. R. D. Smith, The Tellurometer system—new applications to geodesy and hydrography, *J. Geophys. Res.* **65**, No. 2 (1960)
18. Bela Szabó, Geodesy and gravity, Chapter 1, *Handbook of Geophysics and Space Environments*, Air Force Cambridge Research Laboratories, L. G. Hanscom Field, Bedford, Massachusetts.
19. Moody C. Thompson, Jr., A summary of progress in problems related to terrestrial electronic distance measuring in the United States, *Trans. Am. Geophys. Union.* **48**, No. 2 (1967).

20. U. S. Air Force, *Final Report of Results*, Project 53 AFS–1, North Atlantic Tie, 1370th Photo-Mapping Group, Orlando Air Force Base, Florida, 1 March 1958.
21. Robert L. Weaver, Electrotape and the engineer-surveyor, paper presented at the 1965 Convention of the American Society of Photogrammetry/American Congress on Surveying and Mapping, Washington, D.C., March 29–April 2, 1965.
22. William H. Young, Electrotape applications in photogrammetry, paper presented to the Riverside County Flood and Water Conservation District, Riverside, California.

Chapter 8   Gravity

Geometric measurements on the earth's surface—distance and direction—must be reduced to the corresponding quantities on the reference ellipsoid. Computations of position on the ellipsoid were discussed in Chapters 3 and 4, as well as the reduction of observations to the ellipsoid. Relating measurements on the earth to the ellipsoid implies knowledge of the difference between these bodies. In other words, the size and shape of the earth in terms of its departures from a mathematically defined figure must be specified.

Through study of the earth's gravity and its potential, the departures of the earth from the reference ellipsoid may be defined, the center of the ellipsoid may be placed at the center of the earth's gravity, and the shape (flattening) of the ellipsoid may be determined. The size of the ellipsoid, or the value of the equatorial radius, a, cannot be determined from gravity alone. This dimension is obtained through extension of the basic method used by Eratosthenes around 220 B.C. and (or) sophisticated concepts involving our satellite, the moon.

In comparison with the basic geometric considerations discussed thus far, this application of gravity to our endeavor falls in a realm designated "physical geodesy." To explain this aspect of geodesy, expressions for gravity and potential of a rotating sphere will be developed. The terms thus defined then are used to describe the shape of a reference ellipsoid. Expressions showing the geometric deviation of the earth from the ellipsoid as functions of the gravity variation between these surfaces will be cited. Last, the methods of measuring gravity on the earth and the theories of reducing these values to those on a surface defined by mean sea level (the geoid) are discussed.

It is interesting to note that the fundamental mathematics upon which this exercise is based were derived during the 18th and 19th centuries. These forms, however, were not employed extensively for the geodetic objective until the past 50 years because of the sparsity of surface gravity measurements.

GRAVITY AND ITS POTENTIAL

The first step toward understanding is to consider "gravitation." To define gravitation, Newton's universal gravitation formula,

$$F = \frac{Gm_1 m_2}{r^2},$$

is equated with Newton's second law,

$$F = m_1 a,$$

to give

$$-F = \frac{Gm_2}{r^2} = a, \tag{8-1}$$

where F = force (the negative sign is explained subsequently), and G is the Newtonian gravitation constant $(6.673 \times 10^{-8}$ cm^3/gm sec$^2)$. Gravitation, F, is the acceleration, a, experienced at one mass, m_1, due to another mass, m_2, when they are separated by the distance r. This quantity has magnitude as well as direction: it is a vector. The force is directed along r from the attracted body to that doing the attracting. Henceforth, m_1 is considered a unit mass and is set equal to 1, and m_2 is designated M.

The potential of gravitation, V, is a function whose first derivative is gravitation; that is, if

$$V = \frac{GM}{r}, \tag{8-2}$$

then

$$\frac{dV}{dr} = -\frac{GM}{r^2} = F.$$

As r increases, V decreases; this fact leads to the convention of a negative sign for F as shown in Eq. (8-1). This convention is ignored with the acceleration, a. Multiplying Eq. (8-1) by r gives

$$-Fr = \frac{GM}{r} = V,$$

and it is seen that potential is force times distance or work. Potential is defined as the work done by the force field that results from M to bring m_1 from infinity to a point distance r from M. Potential is a scalar value, having magnitude only.

To extend the foregoing concept of point masses to a condition approximating that of the earth, consider a sphere of uniform mass, M, and radius R. In Fig. 8-1, ρ is the distance from m_1 to the center of the sphere.

From Eq. (8-2) for mass points,

$$dV = \frac{G \, dm}{r}$$

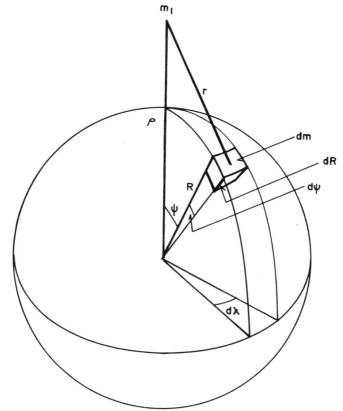

Figure 8–1. Gravitation of a sphere.

is obtained by differentiation of the potential with respect to mass. Likewise, by integration,

$$V = G \int \frac{dm}{r};$$

where dm is a mass element on, or in, the sphere and r is the distance to it. Referring to Fig. 8–1 and designating the density of this sphere by p, note that

$$dm = pR\, d\psi\, R\sin\psi\, d\lambda\, dR = pR^2 \sin\psi\, d\psi\, d\lambda\, dR.$$

Substituting this in the expression for V yields

$$V = Gp \int_0^\pi \int_0^{2\pi} \int_0^R \frac{R^2}{r} \sin\psi\, d\psi\, d\lambda\, dR.$$

From the law of cosines, note that

$$r^2 = R^2 + \rho^2 - 2R\rho\cos\psi;$$

and, further, by differentiating with respect to r and ψ and multiplying the result by R,

$$\frac{R\, dr}{\rho} = \frac{R^2}{r} \sin \psi \, d\psi.$$

Substituting this expression in the foregoing potential integral and writing integration limits for dr in place of $d\psi$ produces

$$V = \frac{G\rho}{\rho} \int_0^{2\pi} \int_{\rho-R}^{\rho+R} \int_0^R R\, d\lambda \, dr \, dR,$$

$$V = \frac{G\rho\, 2\pi}{\rho} \int_{\rho-R}^{\rho+R} \int_0^R R\, dr \, dR,$$

and

$$V = \frac{G\rho\, 4\pi}{\rho} \int_0^R R^2 \, dR = \frac{G\rho\, 4\pi\, R^3}{3\rho}.$$

Since the volume of a sphere is $4/3\,\pi\,R^3$, and its mass, M, is equal to its volume times its density, p,

$$M = p\, 4/3\,\pi\,R^3;$$

therefore,

$$V = \frac{GM}{\rho} \tag{8-3}$$

for a point external to a sphere; that is,

$$\rho \geq R.$$

The gravitation for the sphere is obtained by differentiating Eq. (8–3), that is,

$$\frac{dV}{d\rho} = F = -\frac{GM}{\rho^2}. \tag{8-4}$$

The equation form is the same for the case of the sphere and the point mass; therefore, the mass of a uniform sphere apparently may be considered concentrated at its center.

From Eq. (8–3), it is obvious that V will be constant if ρ does not vary. This condition defines an *equipotential surface*, a surface with a constant potential. For the case of a uniform, nonrotating spherical mass, the surfaces are concentric spheres as one might expect. On these surfaces,

$$\frac{dV}{d\rho} = F = 0$$

because V is constant. Of course there is gravitation, F, on this surface;

therefore, this equation is interpreted to mean that the force of gravitation is in the direction of ρ and is zero along any direction perpendicular to it.

The next step toward the real world is to spin the sphere. The point acted upon is assumed on the sphere, or $\rho = R$.

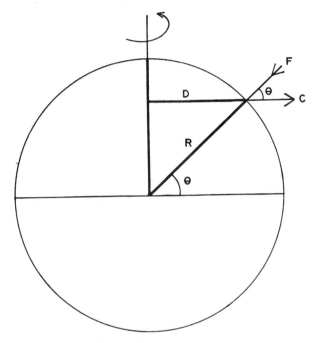

Figure 8–2. Centrifugal force plus gravitation.

Referring to Fig. 8–2, centrifugal force is defined as

$$C = m_1\omega^2 D = \omega^2 R \cos\theta, \qquad (8\text{–}5)$$

where m_1 is the mass of the point acted upon and is given the value of 1, D is the rotation arm and is equal to $R \cos\theta$, and ω is the angular velocity. This force is directed outward, perpendicular to the rotation axis. From Eq. (8–5), it is apparent that centrifugal force at the pole is

$$C_p = \omega^2 R \cos 90° = 0,$$

and this force at the equator is

$$C_e = \omega^2 R \cos 0° = \omega^2 R.$$

The potential of C is

$$V' = \frac{\omega^2 D^2}{2} = \frac{\omega^2 R^2 \cos^2\theta}{2}. \qquad (8\text{–}6)$$

For verification,

$$\frac{dV'}{dD} = \omega^2 D = \omega^2 R \cos\theta = C.$$

Now *gravity* due to the rotating sphere of mass M is the force resulting from vector addition of gravitation and centrifugal force:

$$g = F + C. \tag{8-7}$$

(Recall that F and C are of opposite sign.) Likewise, the potential of gravity is the scalar sum of the potentials of gravitation and centrifugal force:

$$U = V + V' = \frac{GM}{R} + \frac{\omega^2 R^2 \cos^2\theta}{2}. \tag{8-8}$$

Simple differentiation of U with respect to R does not yield a correct vector expression for g in terms of R and $\cos\theta$. An approximate expression for g at any θ is derived as follows (only vector magnitude and not direction is considered in the summations):

At the equator, F and C are along the same line in opposite directions and, therefore, simply may be added,

$$g_e = F + C_e. \tag{8-9a}$$

Likewise, at the pole, as $C_p = 0$,

$$g_p = F. \tag{8-9b}$$

Of course,

$$g_p - g_e = -C_e. \tag{8-9c}$$

At any angle θ, the element of C along the vector F is

$$(\omega^2 R \cos\theta)\cos\theta = C_e \cos^2\theta.$$

Therefore, by combining this element with F and noting Eq. (8–7),

$$g_\theta = F + C_e \cos^2\theta;$$

and, from Eqs. (8–9),

$$g_\theta = g_e - C_e + C_e \cos^2\theta = g_e - C_e \sin^2\theta,$$
$$g_\theta = g_e + (g_p - g_e)\sin^2\theta,$$
$$g_\theta = g_e \left[1 + \left(\frac{g_p - g_e}{g_e}\right)\sin^2\theta\right]. \tag{8-10}$$

This is the expression for the magnitude of gravity experienced at a point on a rotating sphere of uniform mass as a function of angle θ.

Returning to Eq. (8–8), U is seen to be a function of variable R and θ. For an equipotential surface, U must be constant and, because θ varies, so must R. The value of R for $\theta = 0°$ must be larger than that for $\theta = 90°$. Thus, this is the equation of a geometric figure which is nearly an ellipsoid of revolution.

In practice, this ellipsoid is considered to represent the figure that would be assumed by the earth's mean sea level (the geoid) if the earth were a uniform mass. To a first approximation, the ellipsoid flattening in terms of gravity at the pole and equator is

$$f = \frac{g_p - g_e}{g_e},$$

or, as expressed by Clairaut in 1743,

$$f = \frac{5}{2}\frac{C_e}{g_e} - \frac{g_p - g_e}{g_e}. \tag{8-11}$$

To adequately state the potential of an ellipsoid of revolution in place of the sphere, Eq. (8–8) is replaced with

$$U = \frac{GM}{R'} + \frac{Gqa^2}{2R'^3}(1 - 3\sin^2\theta') + \frac{\omega^2}{2} R'^2 \cos^2\theta', \tag{8-12}$$

where R' is the geocentric ellipsoid radius (which varies from a to b), a is the semimajor axis, q is a small term of the order of flattening, and θ' is the geocentric latitude. (Note that this is a different symbol from that defined in Chapters 2 and 6.)

Likewise, the expression for gravity on the ellipsoid is

$$g_\phi = g_e(1 + \beta \sin^2\phi - \beta_1 \sin^2 2\phi), \tag{8-13}$$

where $\beta = (g_p - g_e)/g_e$, $\beta_1 = 0.000006$, and $\phi = $ the geodetic latitude. Notice that Eq. (8–13) is merely an extension of (8–10) for a rotating sphere.

Equations (8–11), (8–12), and (8–13) are abbreviations of series which define a reference figure to which the actual surface is compared. This reference surface is one of specified dimensions and uniform density throughout. Through Eq. (8–11), the value of flattening adopted for the reference ellipsoid may be made consistent with values of g_e, g_p, and C_e observed on the earth. Similarly, determined values of β and β_1 in Eq. (8–13) permit computation of gravity on the reference surface. This is called "normal gravity" and is designated γ. For the International Ellipsoid, Eq. (8–13) becomes

$$\gamma = 978.0490 \ (1 + 0.0052884 \sin^2\phi - 0.0000059 \sin^2 2\phi) \quad \text{cm/sec}^2. \tag{8-14}$$

This is the International Gravity Formula which was adopted by the

General Assembly of the International Union of Geodesy and Geophysics in 1930. Any change in the gravity system (for instance, change in the value assumed for Potsdam) or reference ellipsoid will obviously change the constants in the equation for normal gravity. The values of β and C_e are clearly dependent on values of f and a. Different normal gravity formulas may be derived for different ellipsoids and vice versa. This subject is discussed in [2]. At the present time, published gravity anomalies (the difference between measured and normal gravity) usually refer to the International Ellipsoid.

Gravity values are expressed in terms called the gal, after Galileo Galilei:

$$1 \text{ gal} = 1 \text{ cm/sec}^2; \quad 1 \text{ milligal (mgal)} = 0.001 \text{ gal.}$$

The constants inherent in the International Gravity Formula (IGF), specifically, and others in general terms are:

$$
\begin{aligned}
a \text{ (IGF)} &= 6{,}378{,}388.0 \text{ meters,} \\
f \text{ (IGF)} &= 1/297 = 0.003367003367, \\
\omega &= 0.7292115146 \times 10^{-4} \text{ rad/sec,} \\
C_e \text{ (IGF)} &= 3.391704203 \text{ cm/sec}^2 \text{ (gal),} \\
\gamma_e \text{ (IGF)} &= 978.049000 \text{ gal,} \\
G &= 6.673 \times 10^{-8} \text{ cm}^3/\text{gm sec}^2, \\
M &= 5.973 \times 10^{27} \text{ gm.}
\end{aligned}
$$

From these values, note that $C_e - C_p = 3.392$ gal. From Eq. (8–14), $\gamma_p = 983.221$ gal and, therefore, $\gamma_p - \gamma_e = 5.172$ gal. The reason for the difference between computed gravity at the pole and equator and the difference attributable solely to centrifugal force is the result of ellipsoidal flattening and amounts to 1.780 gal. The pole is in fact nearer the earth's center of gravity (CG) than a point on the equator.

Additionally, because the earth is not of uniform density throughout but increases in density toward the center, the equipotential surface is not really an ellipsoid. The surface, dubbed a spheroid, differs only slightly from an ellipsoid tangent to it at the pole and equator. At latitude 45°, the spheroid is approximately 4.3 meters below the ellipsoid [1]. For this reason, the ellipsoid is accepted as an adequate substitute.

GEOID UNDULATIONS AND DEFLECTIONS OF THE VERTICAL

The ellipsoid of revolution is assumed to have the same volume as the geoid and the same mass as the earth. As noted previously, its center is assumed to coincide with the earth's center of gravity. Equations (8–12) and (8–13) define the potential and gravity, respectively, of this ellipsoid. The potential is equal to that of the geoid—the surface defined by mean sea level theoretically extended under the land masses. Unfortunately, the mass distribution of the earth is not uniform (particularly in the outer shell, the mantle) and, therefore, gravity computed from Eq. (8–13) differs from that determined on the geoid. This difference is a gravity anomaly.

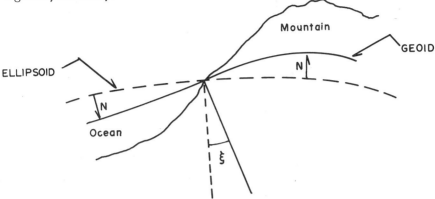

Figure 8–3. Result of mass anomalies. Mass excess corresponds to positive gravity anomaly and positive geoid height.

Figure 8–3 indicates the result of deviation from a uniform mass. The actual mass excesses and deficiencies (mass anomalies) cause the geoid surface to vary from that of the ellipsoid. In the case of mass excess, the geoid is above the ellipsoid. The plumbline of an instrument which defines the vertical to the geoid is diverted from the ellipsoid normal and toward the mass excess. In the case of mass deficiency, the geoid is below the ellipsoid. The basis for the gravimetric theory is that the deflections of the vertical, geoid warp (geoid undulation), and gravity anomalies all result from the same cause—mass anomalies. The gravity anomalies can be observed and then employed to compute the geometric deviation of the geoid from the ellipsoid.

Symbols used for this discussion are:

U = potential of the ellipsoid,
W = potential of the geoid,
γ = gravity on the ellipsoid, *normal gravity*,
g = gravity on the geoid,
Δg = $g - \gamma$ = gravity anomaly, which is truly a gravitational anomaly as the effect of the centrifugal force cancels out in the subtraction.

The condition that the potentials of the ellipsoid and geoid are equal is written

$$U = W = Uw + T = \text{constant},$$

where Uw is the normal potential of the geoid (the potential if there were no mass anomalies), and T is the disturbing potential. Although $U = W$, the two equipotential surfaces do not necessarily coincide and may be separated by N, the geoid undulation or height. (Do not confuse this N with that representing the ellipsoid radius in the prime vertical.) The potential difference between corresponding points on the two surfaces is expressed as

$$T = \int_{R'}^{R'+N} \gamma \, dR',$$

where $dR' = N$, and R' is the ellipsoidal geocentric radius. With N relatively small compared to R', note that

$$T \doteq \gamma N \qquad \text{and} \qquad N \doteq \frac{T}{\gamma}. \tag{8-15}$$

This approximation leads to a working equation with which N may be computed if an expression for T can be found.

An expression for T was derived by Sir G. G. Stokes. The derivation is beyond the intended scope of this book. The Stokes function is

$$N_g = \frac{1}{4\pi\gamma_m} \int_0^{2\pi} \int_0^{\pi} \Delta g\,(\psi, \alpha)\, S(\psi) \sin \psi \, d\psi \, d\alpha; \tag{8-16}$$

where γ_m is the mean value of normal gravity; ψ is the angular distance between the point where N is being determined (m_1) and the area where the effect of Δg is being considered (see Fig. 8–1); α is the azimuth from the affected point (m_1) to that causing the effect (dm);

Δg is the mean gravity anomaly of the affecting area; and

$$S(\psi) = \csc \frac{\psi}{2} + 1 - 6 \sin \frac{\psi}{2} - 5 \cos \psi - 3 \cos \psi \ln\left(\sin \frac{\psi}{2} + \sin^2 \frac{\psi}{2}\right). \tag{8-17}$$

(Subscript g indicates gravimetric or absolute values.)

Integration should be carried out over the entire surface. The major contributions to N_g, however, are due to values of Δg within a few hundred miles of the point; therefore, the effects of distant areas are approximated with mean anomalies over large areas. In practice, this computation is accomplished with a grid similar to that shown in Fig. 8–4. Mean gravity anomalies are estimated for areas encompassed by azimuth lines and distance rings. N is a function of the distance ψ to dm and its gravity anomaly, Δg. Such templates take different forms, because of the varying assumptions used to simplify the computations [4–6].

It should be noted that Eq. (8–16) is valid only for points on, or

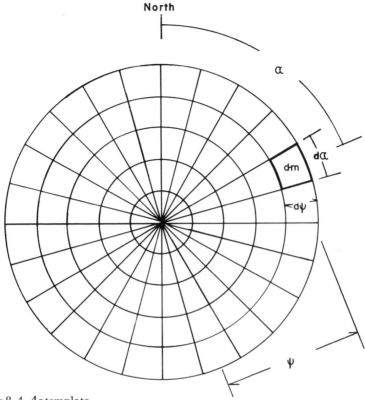

Figure 8–4. Δg template.

exterior to, the equipotential surface—the assumption being that all mass producing this potential is contained within the surface, the geoid.

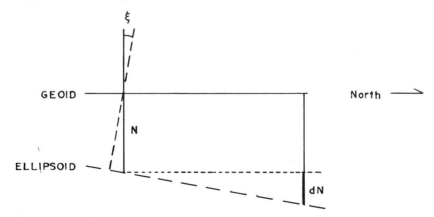

Figure 8–5. Deflection of the vertical.

As may be noted from Figs. 8–3 and 8–5, the angular deviations between the ellipsoid normal and geoid vertical are related to the change in $\mathcal{N}$.

Specifically,

$$\xi_{\mathrm{g}} = -\frac{d\mathcal{N}_{\mathrm{g}}}{dx}, \text{ in the north-south direction, positive north;}$$

$\xi_{\mathrm{g}} = $ the geoidal latitude (astro)–ellipsoid latitude (absolute),

$\xi_{\mathrm{g}} = \varPhi - \phi_g;$

$$\eta_{\mathrm{g}} = -\frac{d\mathcal{N}_{\mathrm{g}}}{dy} \text{ in the east-west direction, positive east,}$$

$\eta_{\mathrm{g}} = (\varLambda - \lambda_g) \cos \phi.$

Differentiating (8–16) gives

$$\xi_{\mathrm{g}}'' = \frac{1}{4\pi\gamma_{\mathrm{m}}\sin 1''} \int_0^\pi \int_0^{2\pi} \varDelta_{\mathrm{g}}(\psi, \alpha) \frac{dS(\psi)}{d\psi} \sin \psi \cos \alpha \, d\psi \, d\alpha, \quad (8\text{–}18)$$

$$\eta_{\mathrm{g}}'' = \frac{1}{4\pi\gamma_{\mathrm{m}}\sin 1''} \int_0^\pi \int_0^{2\pi} \varDelta_{\mathrm{g}}(\psi, \alpha) \frac{dS(\psi)}{d\psi} \sin \psi \sin \alpha \, d\psi \, d\alpha, \quad (8\text{–}19)$$

where

$$\frac{dS(\psi)}{d\psi} \sin \psi = -\frac{1}{\sin(\psi/2)} - 3 - 8 \sin \frac{\psi}{2} + 32 \sin^2 \frac{\psi}{2} + 12 \sin^3 \frac{\psi}{2}$$

$$- 32 \sin^4 \frac{\psi}{2} + 3 \sin^2 \psi \ln\left(\sin^2 \frac{\psi}{2} + \sin \frac{\psi}{2}\right). \quad (8\text{–}20)$$

These expressions are the work of the late Dr. F. A. Vening Meinesz and bear his name.

Gravimetrically computed $\mathcal{N}_g$, ξ_g, and η_g are functions of $\varDelta g$ and, therefore, depend on the normal gravity formula used. Geometrically, this dependence is limited to the flattening of the reference ellipsoid. It is to be emphasized that these departures are relative to an ellipsoid with its center coinciding with the CG of the geoid.

By comparison, the astrogeodetic deflections of the vertical discussed in Chapters 5 and 6 depend on all dimensions of the reference ellipsoid as well as its orientation relative to the earth.

Assuming an adequate value for a has been determined through properly reduced arc measurements and that a value for flattening consistent with gravity is used, geodetic coordinates on an absolute datum can be determined with astronomical coordinates and gravimetric deflections of the vertical; that is,

$$\phi = \varPhi - \xi_g, \quad (8\text{–}21)$$
$$\lambda = \varLambda - \eta_g \sec \phi, \quad (8\text{–}22)$$
$$\alpha = A - \eta_g \tan \phi. \quad (8\text{–}23)$$

In theory, this method could relate all regional data to one global datum. Astronomical coordinates would be converted to geodetic co-ordinates on this global datum when gravimetric deflections have been determined. In essence, this is the objective of the geodesist.

GRAVITY MEASUREMENTS

Now that the purpose of gravity measurements and their benefits have been described, methods of acquiring values for gravity will be discussed. Basically, there are two types of gravity measurements—absolute and relative.

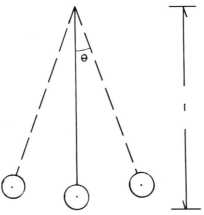

Figure 8–6. Simple pendulum.

Absolute gravity may be determined with a pendulum, the pendulum's period being a function of gravity. The period of a simple pendulum is

$$P = \pi \left(\frac{l}{g}\right)^{1/2} \left(1 + \frac{\theta^2}{16} - \cdots\right),$$

where P is the half-period (that is, the time to swing from one side to the other), l is the length, g is gravity, and θ is the angle of swing. See Fig. 8–6. Because θ is small (usually less than 1°), the last portion will be disregarded. The equation becomes

$$P = \pi \left(\frac{l}{g}\right)^{1/2}, \qquad (8-24)$$

and obviously

$$g = \frac{\pi^2}{P^2} l.$$

There are several errors which affect pendulum measurements:
1. Even though the pendulum is enclosed in an evacuated chamber,

the friction generated by the residual air dampens the pendulum swing. Also, this air tends to support the pendulum by the amount that is displaced and causes a buoyant effect.

2. Molecules of air are adsorbed on the shank of the pendulum and this raises the center of gravity of the pendulum, thereby changing l.

3. There is a sympathetic vibration of the foundation on which the pendulum rests. This can be canceled by swinging a second pendulum in the opposite direction.

The absolute determination of gravity is seldom made except to establish gravity base stations. For relative gravity measurements with a pendulum in the determination of gravity differences between stations, the following equations hold:

$$P_1{}^2 g_1 = \pi^2 l \quad \text{at station 1,}$$
$$P_2{}^2 g_2 = \pi^2 l \quad \text{at station 2.}$$

Because the right hand sides of these equations are identical,

$$P_2{}^2 g_2 = P_1{}^2 g_1;$$
$$g_2 = \frac{P_1{}^2}{P_2{}^2} g_1. \tag{8-25}$$

To determine gravity with accuracy, time must be measured to a high precision. This is demonstrated by differentiating P with respect to g in Eq. (8–24):

$$dP = -\frac{P}{2}\frac{dg}{g}.$$

Assume the following conditions:

$$dg = 1 \text{ mgal}, \qquad g = 10^6 \text{ mgals}, \qquad P = 0.5 \text{ second.}$$

With these values substituted,

$$dP = 2.5 \times 10^{-7} \text{ second.}$$

The negative sign has been removed because time is a positive quantity. This is the precision with which time must be measured in order to determine the value of gravity to ± 1 mgal—hardly a coarse measurement. It is generally conceded, however, that the pendulum method yields accuracies of 1 mgal.

Free-falling bodies as well as the rise and fall of projected bodies have been used experimentally to determine absolute g. Such methods also demand precise time and length fixes.

Absolute gravity stations exist the world over. By agreement of the International Gravity Commission, the value of 981.274 gals at the Pendelsaal in Potsdam is accepted as the international reference. Relative gravity ties from other absolute stations, however, indicate this value may be 13 mgals too high. This discrepancy has little effect on geoid determinations as it applies equally to γ and g and is thereby cancelled in Δg.

Station gravity is usually determined by measuring the gravity difference between the station and a reference station. In practice, gravimeters are employed for these measurements of *relative gravity*.

Nonpendulous gravity meters are referred to as gravimeters. There are many types of gravimeters making use of various principles. The Haalck gravimeter employs the barometric principle; the Boliden, an electric method. However, most designs use the basic idea of a small mass suspended from a responsive element (such as a spring) with displacement being a function of gravity. In this category, there are two types— the stable and the unstable.

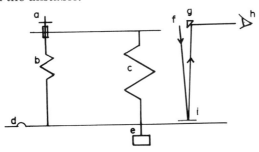

Figure 8–7. Stable gravimeter. a, micrometer; b, adjusting spring; c, main spring; d, hinge; e, mass; f, light source; g, prism; h, eye; i, mirror.

In the stable gravimeters, the responsive element balances gravity with an opposing force. There is a displacement which can be measured, thereby indicating gravity relative to a reference. Figure 8–7 illustrates a Hartley gravimeter. An increase or decrease in gravity will lengthen or shorten the main spring. It can be returned to its fixed reference value by increasing or decreasing the tension of the adjusting spring. The magnitude of the screw adjustment is a direct function of the change in gravity from its reference.

Unstable gravimeters are designed in such a manner that any departure from equilibrium brings a third force into play which magnifies the displacement that would be caused by gravity alone. The operating principle of a Thyssen gravimeter is depicted in Fig. 8–8. In the upper sketch, there is a balance between gravity and the main spring. The auxiliary weight is directly above the pivot and exerts no turning moment on the horizontal beam. In the lower sketch, a change in g has tilted the beam slightly. The auxiliary weight has moved so that it exerts a turning moment on the beam, thus reinforcing the turning

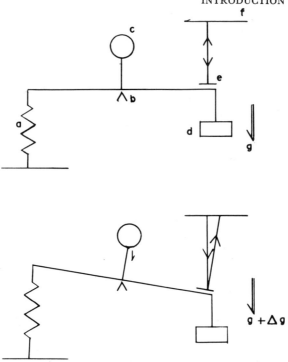

Figure 8–8. Unstable gravimeter. a, balance spring; b, pivot; c, auxiliary weight; d, mass; e, mirror; f, scale; g, gravity; Δg, increase in gravity.

moment instigated by gravity and causing additional spring elongation. Obviously, this principle eases the problem of measuring small displacements. The Worden and the LaCoste–Romberg gravimeters are of this unstable type.

An additional factor is the range through which a gravimeter will operate. Most meters operate through a 50- to 200-mgal range which is not really adequate for the geodetic purpose. It is significant to note that the Worden and the LaCoste–Romberg instruments have the required accuracy and a wide operating range.

Because of their light weight (10 pounds), tough construction, low cost ($8000), accuracy (from 0.1 to 0.01 mgals), and speed of operation, gravimeters logically replace pendulums for measurements of relative gravity.

The foregoing discussion dealt with measurements on land. Because water occupies roughly 80% of our earth, and because gravity over the entire surface should be considered in solving Eqs. (8–16)–(8–20), a brief comment concerning gravity measurements at sea is warranted. The idea of alleviating sympathetic vibrations by swinging pendulums in

Figure 8–9. LaCoste–Romberg gravimeter (official U.S. Air Force photograph).

opposite directions was devised by Vening Meinesz 50 years ago. This concept has been successfully extended to use on board submarines. The transverse and vertical accelerations are not significant, while the longitudinal movements are compensated with this double pendulum apparatus. Accuracy of gravity measurements from submarines varies varies between 2 and 5 mgals.

Gravimeters for use on surface ships have been developed by Anton Graf of Germany and LaCoste–Romberg and Bell Aerosystems of the United States [3]. Test results have varied considerably depending on sea state, techniques, and so on. As a general statement, accuracies within ±7.5 mgals seem possible about 70% of the time.

In off-shore areas, diving bells with gravimeters and operators have been used in depths up to 125 feet, and remotely controlled gravimeters have been employed at depths of 600 feet. In both cases, the accuracies achievable are comparable with those on land.

GRAVITY REDUCTIONS

Gravity measurements made on the earth's surface must be reduced to the geoid for computation of the gravity anomalies employed with

Eqs. (8–16)–(8–20). The reductions must attempt to produce the value of g that would have occurred if it were possible to observe on the geoid surface. There are many such reductions; however, only those major concepts applicable to geodesy are discussed—free-air, Bouguer, and isostatic, plus the terrain correction which may be applied to each.

To be adequate for use with the Stokes function applied to geodesy, any gravity reduction should satisfy the following conditions summarized by:

(1) no masses outside the geoid,
(2) no change of total mass,
(3) no change in locations of the center of gravity, and
(4) no surface deformation to the geoid.

It is important to note that no matter what gravity reductions and resulting anomalies are used, the physical values of undulations and deflections would be the same provided these conditions are met or accounted for.

Free-Air Reduction. This reduction merely takes the gravity observed at elevation H and reduces it to the geoid. If gravity on the geoid is

$$g_0 = \frac{GM}{R^2},$$

then

$$g = \frac{GM}{(R+H)^2}$$

is gravity at elevation H, or, by series expansion,

$$g = \frac{GM}{R^2}\left(1 - \frac{2H}{R} + \frac{3H^2}{R^2} - \cdots\right) = g_0 - \frac{2g_0 H}{R} + \cdots \quad .$$

From this expression, the correction may be written,

$$\delta'_t = g_0 - g = \frac{2g_m H}{R_m}, \qquad (8\text{–}26)$$

where R_m and g_m are the average radius of curvature and gravity values, respectively, and δ_t' is the free-air correction. With average values inserted, the free-air correction for H measured in meters is,

$$\delta'_t = 0.3086H \quad \text{mgal.}$$

For positive elevations on land, this correction is always added to the observed gravity because the closer the attracting mass, the greater the gravity. The effect of the intervening mass on the measured value of g has not been removed; therefore, this mass may be thought of as having been pushed into or spread on the surface of the geoid. The effects of terrain may be accounted for with a terrain correction (this correction is explained later). The free-air reduction, as depicted in Fig. 8–10, combines the free-air correction and terrain correction and is written

$$\delta g_f = \delta'_f + \delta'_t . \qquad (8\text{–}27)$$

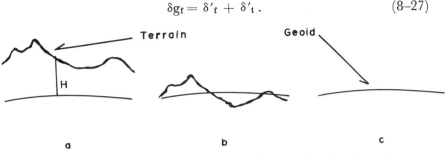

Figure 8–10: Free-air reduction. a, gravity observed at elevation H; b, free-air correction; c, terrain correction.

The change of mass distribution (not total mass) changes the potential at the geoid and, essentially, a new surface is formed a few meters below the geoid. This new surface is called a co-geoid. The mass outside this co-geoid must be placed within it for the same reason that mass exterior to the geoid must be manipulated. This is accomplished by accounting for the "indirect effect." Undulations determined with reference to the co-geoid may be converted to the geoid with knowledge of the distance between these surfaces. The interested reader is referred to [1] for details.

The free-air reduction satisfies conditions 1 and 2. It does change the center of gravity (condition 3) and it does deform the geoid surface by a few meters.

Bouguer Reduction. This reduction moves the mass between the earth's surface and the geoid to infinity and then reduces the point to the geoid.

The first step is to derive an expression for the attraction of a plate of infinite extent that exists between the measured point and the geoid. This step assumes a flat earth. With reference to Fig. 8–11, the potential at m_1 resulting from a mass element, dm, on a disk of thickness dh may be written

$$U = \frac{G\,dm}{r},$$

and

$$dm = dh\, D\, d\alpha\, dD\, p,$$

where p is the density of the disk. Also, $r^2 = D^2 + H^2$, from which we note

$$r\,dr = D\,dD.$$

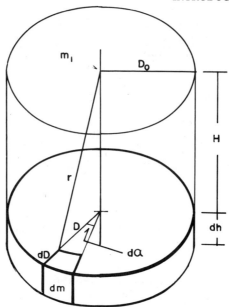

Figure 8–11. Bouguer disk.

Substituting this relation in that for dm yields

$$dm = dh \, r \, dr \, d\alpha \, p.$$

With this expression for dm, the potential expression for the entire disk of thickness dh and radius D_0 becomes

$$U_{dh} = Gp \, dh \int_0^{2\pi} \int_H^{(H^2 + D_0^2)^{1/2}} d\alpha \, dr;$$

$$U_{dh} = Gp \, dh \, 2\pi \, [(H^2 + D_0^2)^{1/2} - H].$$

The attraction of this disk is

$$F_{dh} = \frac{dU_{dh}}{dH} = - Gp2\pi \, dh\left(1 - \frac{H}{(H^2 + D_0^2)^{1/2}}\right).$$

Now the attraction of the entire disk of thickness H is

$$F_H = - Gp2\pi\left[\int_0^H dh - \int_0^H \frac{H \, dh}{(H^2 + D_0^2)^{1/2}}\right],$$

and

$$F_H = - Gp2\pi(H - (H^2 + D_0^2)^{1/2} + D_0). \tag{8–28}$$

For a disk of infinite radius $(D_0 \to \infty)$, H is assumed small relative to D, and this equation may be written

$$F_H = - Gp2\pi H;$$

or, recalling that for a sphere of radius R (disregarding C)

$$g = -\frac{GM}{R^2} \quad \text{and} \quad M = 4/3R^3\pi p_m,$$

and substituting an expression for G derived therefrom, F_H may be written

$$\delta'_B = F_H = \frac{3}{2}\frac{p}{p_m}\frac{H}{R_m}g_m, \tag{8-29}$$

where R_m, p_m, and g_m are the mean values of radius, density, and gravity respectively. This is the Bouguer correction.

With $p = 2.67$ and $p_m = 5.576$, then

$$\delta'_B = +0.1108H \quad \text{mgal},$$

where H is given in meters.

For positive elevations, this correction is subtracted from measured gravity to remove the effect of the mass between the point and the geoid. There are other forms of this correction and the numerical coefficient for H depends on the values of p and p_m used. The reader is referred to [2] for details.

Now that the intervening mass, the Bouguer plate, has been removed, the remaining gravity is reduced to the geoid with the free-air reduction. The final Bouguer reduction takes the form

$$\delta g_B = \delta'_f + \delta'_t - \delta'_B = (0.3086 - 0.1108) \quad \text{mgal/meter},$$

or

$$\delta g_B = 0.1978H \quad \text{mgal}, \tag{8-30}$$

where H is in meters.

Recall that Eq. (8–29) was derived for a flat disk or plate. Obviously, this assumption has shortcomings when the nonflat earth is being considered. Therefore, an additional consideration for curvature is applied in the Bouguer reduction for areas 17 km or farther from the station. These corrections are tabulated in [2].

The effect of δg_B is to remove the mass spread on the geoid surface with δg_t.

The Bouguer reduction only satisfies condition 1, that is, no masses outside the geoid. To a considerable degree, it changes the earth's mass and the center of gravity, and deforms the geoid by hundreds of meters. Its primary value is in geophysical surveys when it is desired to locate mass anomalies. Though its direct application to geodesy is scant, it is employed in conjunction with isostatic reductions and its derivation provides concepts to be applied in other reductions.

Isostatic Reduction. Before this reduction is discussed, the theory supporting it must be explored. This theory not only provides an important tool to the geodesist, but also sheds considerable light on the structure of our earth.

The origin of isostasy (which literally means "equal standing") is to be found in India about the middle of the nineteenth century. About 1850 in India a precise survey was conducted that included the towns of Kaliana in the north and Kalianpur to the south. The distance between these towns, about 375 miles, was measured by the most precise survey methods of the time. An accuracy of a few meters was expected. However, the position of Kaliana as determined from the survey differed by 5 arc seconds, approximately 150 meters, from that determined astronomically.

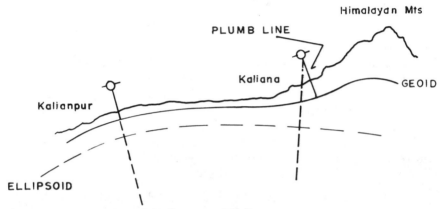

Figure 8–12. Section through Kalianpur and Kaliana.

Archdeacon J. H. Pratt attempted to explain this discrepancy by hypothesizing that the mass of the Himalayas deflected the plumbline more to the north at Kaliana than at Kalianpur. Figure 8–12 illustrates the problem. Remember, astronomical latitude would be based on the plumbline, not the normal to the reference ellipsoid. After scaling the mass of the Himalayas, Pratt found their effect to be 15 seconds and not the 5 seconds actually measured. He could not immediately explain this discrepancy.

G. B. Airy (the Astronomer Royal) submitted an explanation in 1855 that has come to be known as the "roots-of-the-mountains" concept. He reasoned that the mass of the Himalayas sank into the heavier material of the mantle until it received sufficient buoyancy to float. The light material in the mountain roots would reduce the northward attraction calculated by Pratt. In the case of the sea, Airy postulated the existence of heavy antiroots beneath the ocean depths. The lighter ocean water would tend to cancel the attraction of the heavy antiroots. Seismic evidence generally supports the Airy theory.

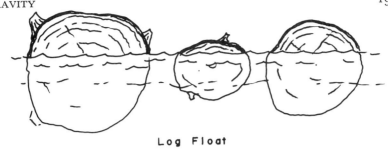

Log Float

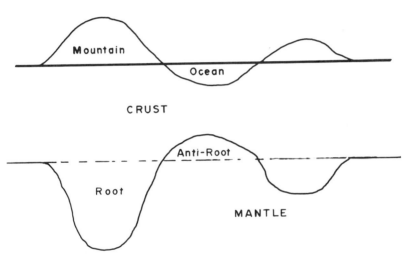

Figure 8–13. Airy isostatic theory.

This theory finds an analogy in a raft of logs floating in the water. If the density of all logs is the same, a large log not only will rise higher in the water than will a small log, but its depth below the surface will be greater. Each log will sink until it displaces enough water to float. Figure 8–13 depicts this theory.

If mountains, valleys, and ocean basins are compensated as Airy theorized, the isostatic gravity anomaly based on this concept will be zero. On the other hand, in the case of a newly formed volcano which had not adjusted in the mantle, the gravity anomaly would be greater than zero; in the case of an eroded mountain with its root remaining, the gravity anomaly would be less than zero.

Four years later, in 1859, Pratt proposed a different explanation. He reasoned that at a certain depth beneath the surface there exists a level of constant pressure produced by the overlaying material—a base that supports a uniform weight per unit of area. To accommodate this theory, he believed that within and under mountains there was a deficiency of density, possibly due to expansion of material caused by heating; on the other hand, under oceans the density would be greater than average to compensate for the less than average density of water.

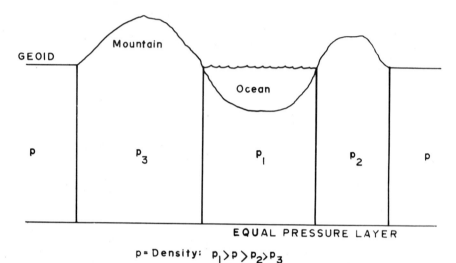

Figure 8–14. Pratt isostatic theory.

The Pratt theory is familiarly referred to as the "dough hypothesis." Assume that equal weights of dough have been placed in identical pans and that these pans are then exposed to different baking conditions. The dough in one pan would rise higher than that in another depending on conditions; however, the pressure on the bottom of each pan would remain constant as the weight of the dough had not changed. Figure 8–14 illustrates the Pratt concept.

From these concepts have come two working theories:

Pratt-Hayford Theory. J. F. Hayford proposed that the depth of compensation (depth of equal pressure) be measured from the topography rather than from the geoid. Though this value came from trial and error, this depth produces the smallest residual gravity anomalies within the United States.

Airy-Heiskanen Theory. Dr. W. A. Heiskanen popularized the Airy theory and concluded that the normal thickness of the earth's crust (no roots or antiroots) is close to 30 km.

Now that the theory has been explored, its application will be explained. The discussion is centered on the Airy-Heiskanen approach, which most closely represents the true condition. The concepts developed generally apply to the Pratt thesis (the outstanding exception being the use and determination of t, the thickness of the mountain root).

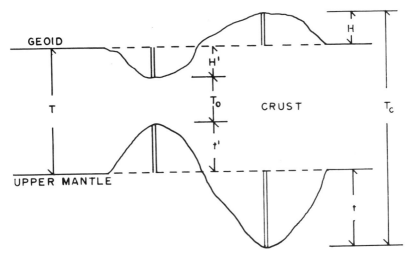

Figure 8–15. Airy–Heiskanen isostatic theory.

First, the mathematics necessary to determine t are confronted. With reference to Fig. 8–15, the symbols are defined as follows:

$$
\begin{aligned}
T &= \text{average crustal thickness} = 30 \text{ km},\\
H &= \text{mountain elevation},\\
t &= \text{thickness of the mountain root},\\
T_c &= \text{crustal thickness of the mountain section},\\
H' &= \text{ocean depth},\\
t' &= \text{thickness of the antiroot},\\
T_o &= \text{crustal thickness of the ocean section},\\
p &= \text{average crustal density} = 2.67,\\
p' &= \text{density of the upper mantle} = 3.27.
\end{aligned}
$$

Assume that a small column of material in the mountain is balanced by another column in the mountain root. The two columns will be in equilibrium under the following conditions:

$$Hp = t(p' - p);$$

therefore,

$$t = \frac{p}{p' - p} H = \frac{2.67}{0.60} H = 4.45H,$$

the thickness of the root. The total crustal thickness at the mountain section becomes

$$T_c = T + H + t = T + H + 4.45H = T + 5.45H. \qquad (8\text{–}31)$$

For the case of the ocean, the procedure is similar. Assume that a column of deficient mass in the ocean basin is balanced by a column of excess mass in the antiroot, that is,

$$t'(p' - p) = H'(p - 1.03),$$

1.03 being the ocean's density. Likewise,

$$t' = \frac{1.64}{0.6} H' = 2.73\, H',$$

is the thickness of the antiroot. The total thickness of the ocean section becomes,

$$T_0 = T - H' - t' = .T - 3.73\, H'. \qquad (8\text{–}32)$$

The next step is to express the vertical attraction equivalent to the isostatic compensation. An equation may be derived by modifying Eq. (8–28) developed in conjunction with the Bouguer correction; specifically,

$$F = -Gp2\pi(H - [H^2 + D_0^2]^{1/2} + D_0),$$

which expresses the attraction of a disk of radius D_0, thickness H and density p. The attraction of interest is that due to a disk of thickness t and density Δp, and located at a distance T below the geoid. If Δp is substituted for p, T^2 for H^2 in the radical, and $(D_0^2 + (T + t)^2)^{1/2}$ for $H + D_0$, then

$$F_i = -G\Delta p2\pi[\,(D_0^2 + (T + t)^2)^{1/2} - (D_0^2 + T^2)^{1/2}\,].$$

To accommodate this correction over the entire earth, the disk is partitioned into rings (zones) of widths $D_2 - D_1$. Then,

$$\begin{aligned}F_{ic} = -\,G\Delta p2\pi[\,&(D_1^2 + (T + t)^2)^{1/2} - (D_1^2 + T^2)^{1/2} \\ &- (D_2^2 + (T + t)^2)^{1/2} + (D_2^2 + T^2)^{1/2}\,]; \qquad (8\text{–}33)\end{aligned}$$

in effect, the contribution of the disk (cylinder, actually) of radius D_1 is subtracted from that of D_2. For oceans, this equation is written

$$\begin{aligned}F_{io} = G\Delta p2\pi[\,&(D_1^2 + T^2)^{1/2} - (D_1^2 + (T - t')^2)^{1/2} \\ &+ (D_2^2 + (T - t')^2)^{1/2} - (D_2^2 + T^2)^{1/2}]. \qquad (8\text{–}34)\end{aligned}$$

As D increases, it is necessary to divide the rings into n compartments and the foregoing equations then must also be divided by n.

These equations comprise the isostatic correction,

$$\delta'_i = F_{ic} \text{ or } F_{io}.$$

This correction is achieved by drawing rings of increasing radii around the computation point and then dividing these rings into compartments (Hayford Zones and Compartments are shown in [2] and are used for this purpose.) This template is similar to Fig. 8–4. From topographic maps, the mean elevation for each compartment is obtained and is used to compute T_c or T_o from Eq. (8–31) or (8–32), and then δ_i from Eq. (8–33) or (8–34).

Heiskanen has devised tables keyed to the Hayford Zones and Compartments, with values for $T = 20, 30,$ and 40 km. With mean elevation, these tables yield a coefficient point for each zone, due to the isostatic as well as the topographic corrections. These effects are considered over the entire world by summing the results derived for each compartment.

The final isostatic reduction combines the corrections for free-air, topography, Bouguer, and isostasy, that is,

$$\delta g_i = \delta'_f + \delta'_t - \delta'_B + \delta'_i. \tag{8–35}$$

The isostatic correction replaces the mass removed by the Bouguer correction and replaces it in a manner consistent with the actual condition on the earth. The isostatic reduction satisfies conditions 1 and 2— no mass outside the geoid and no change of total mass—and nearly satisfies condition 3, change of the CG. It can deform the geoid in certain situations.

The topographic correction (also called terrain) referred to in the foregoing reductions considered the effect of land which is higher and lower than the gravity station (m_1). Refer to Fig. 8–10. In the case of the hill, the material is higher than the station and exerts an upward component of attraction which opposes the downward pull of gravity. A correction must be added to the observed gravity to remove this effect. In the Bouguer correction, the effect of the mass in the valley below the station was subtracted with the plate. Because this material is actually missing, we must add a correction to restore what the Bouguer correction erroneously removed. Thus, for the case of both the hill and the valley, *corrections must be added.*

The attraction of the terrain may be expressed by referring again to Eq. (8–28). If H is the elevation of the point being reduced, if H_T is the mean elevation of the compartment (Hayford's), and if the width of the ring or zone is $D_2 - D_1$ and there are n compartments, the vertical attraction of the compartment terrain on the computation point is

$$\delta'_t = F_t = - \frac{Gp2\pi}{n} [(D_1{}^2 + (H + H_T)^2)^{1/2} - (D_1{}^2 + H^2)^{1/2}$$
$$- (D_2{}^2 + (H + H_T)^2)^{1/2} + (D_2{}^2 + H^2)^{1/2}].$$

As may be visualized, this could be a tedious process. The procedure followed is identical to that of the isostatic correction and this topo-graphic effect is included in the tables mentioned.

For the case of observations on land, the gravity corrections, re-ductions, and the resulting anomalies are summarized:

$$\text{Free-air anomaly:} \quad \Delta g_\mathrm{f} = g + \delta'_\mathrm{f} + \delta'_\mathrm{t} - \gamma, \qquad (8\text{--}36)$$
$$\text{Bouguer anomaly:} \quad \Delta g_\mathrm{B} = g + \delta'_\mathrm{f} + \delta'_\mathrm{t} - \delta'_\mathrm{B} - \gamma,$$
$$\text{Isostatic anomaly:}^* \; \Delta g_\mathrm{i} = g + \delta'_\mathrm{f} + \delta'_\mathrm{t} - \delta'_\mathrm{B} + \delta'_\mathrm{i} - \gamma.$$

*Type (Pratt-Hayford or Airy-Heiskanen) and compensation depth assumption should be noted.

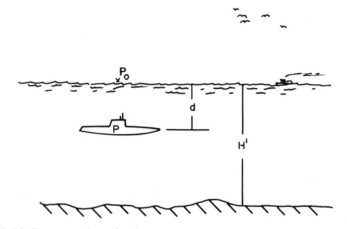

Figure 8–16. Ocean gravity reduction.

To this point, observations of gravity on land (a positive elevation) have been the object of discussion. Now, what happens with gravity observations made aboard a submarine? Actually, gravity reductions for observations at sea are quite similar to those made on land. Figure 8–16 indicates a submarine cruising at depth d in an ocean H' meters deep. The gravity meter is at P directly beneath the surface point, P_0, where the magnitude of gravity is desired.

The *free-air correction*, as on land, accounts for the position of the gravity meter with respect to sea level. It is closer to the earth's center of gravity at P than it would have been at P_0; therefore, the correction must be subtracted.

In this case,

$$\delta'_\mathrm{f} = \frac{2 g_\mathrm{m}}{R_\mathrm{m}} \, d = 0.3086 d \quad \text{mgal},$$

where d is given in meters. By itself, the correction has no meaning, as explained in the next paragraph.

The *Bouguer plate correction* accounts for the thin plate of water of infinite extent which is above the submarine. This correction must be applied twice, once to remove the effect of the upward pull of the plate of water, and again to restore the effect of the downward pull of the plate of water if the gravity meter had been on the surface. The reduction is always positive.

$$2\delta'_B = 2\left(\frac{3}{2}\frac{p}{p_m}\frac{d}{R_m} g_m\right) = 0.0861d \quad \text{mgal,}$$

where d is given in meters; $p = 1.03$, the density of sea water; and $p_m = 5.576$, the mean density of the earth. This correction is combined with the foregoing free-air correction to result in an anomaly the authors will designate "free-air" for observations made at sea.

There is a *second Bouguer correction*. This correction accounts for the effect of the slab of water of thickness H' which extends to the ocean floor. This correction is based on the assumption that the ocean basins represent a mass deficiency not compensated isostatically. It is necessary, according to Bouguer, to remove the mountains completely (which was done with Bouguer correction on land) and, similarly, to carry additional masses to the ocean basin in order to fill them with material of density 2.67. Because the oceans are already filled with material of density 1.03, it is necessary to consider only the density difference $(2.67 - 1.03 = 1.64)$. This correction is always positive.

$$\delta'_{B2} = \frac{3}{2}\frac{(p - 1.03)}{p_m}\frac{H'}{R_m} g_m = 0.0688H' \quad \text{mgal,}$$

where H' is expressed in meters.

The *isostatic correction* is quite similar to the land case. The thickness, T_o, of the crust under the ocean can be computed with Eq. (8–32). With Eq. (8–34) keyed to appropriately charted zones and compartments, the isostatic correction is computed.

The last correction to be presented is of a different type and is unique to gravity observations made from a moving vehicle. The *Eötvös correction* (after the late Hungarian physicist, Baron Roland von Eötvös) is necessary to account for the east–west component of the velocity of the submarine. This component of travel either increases (from west to east) or decreases centrifugal force, and therefore, must be removed from observations. This correction is expressed as

$$\delta'_E = \pm 2\omega v \cos\phi = \pm 0.14584v \cos\phi \quad \text{mgal,}$$

where v is the east-west velocity component in centimeters/second, ϕ is the geodetic latitude, and ω is the rotational velocity of the earth, 0.7292×10^{-4} radian/second. The Eötvös correction is always positive for eastward travel and negative for westward travel.

Finally, there is the requirement to know the latitude of the submarine to a relatively high degree of accuracy. Recall that gravity does vary with latitude according to the International Gravity Formula, Eq. (8–14). The rate of change of gravity along a north–south line is

$$0.81 \sin 2 \phi \quad \text{mgal/km,}$$

and at latitude 45°, the gravity variation is about 0.1 mgal for each 120 meters traveled in a north-south direction. This means that in order to obtain an anomaly accuracy of 1 mgal, the position must be known to at least 1200 meters.

As was done previously, the gravity corrections for *observations made at sea* are recapitulated by listing the resulting gravity anomalies:

Free-air anomaly: $\quad \Delta g_f = g - \delta'_f + 2\delta'_B \pm \delta'_E - \gamma,$

Bouguer anomaly: $\quad \Delta g_B = g - \delta'_f + 2\delta'_B + \delta'_{B2} \pm \delta'_E - \gamma,$

Isostatic anomaly: $\quad \Delta g_i = g - \delta'_f + 2\delta'_B + \delta'_{B2} \pm \delta'_E - \delta'_i - \gamma.$

The philosophy is identical for the reduction of gravity observed from an aircraft. In essence, the corrections may be summarized as follows.

Free-air correction: same as on land.

Bouguer correction: same as on land except that the Bouguer slab is remote from the observing station.

Isostatic correction: same as on land.

Topographic correction: same as on land. This correction is zero over the ocean.

Eötvös correction: same as the case at sea except that, due to high aircraft velocities, the accuracy of the velocity determination must be higher.

To illustrate the magnitude of the anomalies resulting from these reductions under different observation conditions, examples are shown in Table 8–1.

It is apparent that the Bouguer anomalies are negative on land and positive in oceans, and are generally much larger than either free-air or isostatic anomalies. Assuming that the isostatic correction is a valid representation of the truth, this result is to be expected. If there is, in fact, a root of lighter material extending under mountains, then the measured value of gravity on the mountain top would be less than if the mountain were merely excess material piled on top of a crust of uniform density and thickness. Because the Bouguer correction does not consider this condition, the less-than-anticipated value of gravity is reduced uncompensated and results in a negative anomaly. The opposite reasoning may be applied to the situation in ocean areas.

In keeping with the condition that mass must not be discarded or

TABLE 8-1

Gravity Anomalies in Milligals [2]

	H (meters)	Δg_f	Δg_B	Δg_i (Airy-Heiskanen $T=40$ km)
Mountain Terrain				
Las Vegas	1960	+36	−181	+8
Truckee	1805	+45	−154	−4
Grand Canyon	2386	+52	−200	0
Lake Placid	571	+38	−9	+16
Valley Terrain				
Knoxville	280	−6	−37	−9
El Paso	1146	+24	−103	+29
Colorado Springs	1841	+2	−180	+8
Salt Lake City	1322	−15	−138	+26
Sea				($T=30$ km)
Mediterranean	−2200	−1	+149	+8
Indian Ocean	−4390	−33	+266	−30
North Atlantic Ocean	−3610	+26	+272	+36
South Atlantic Ocean	−3700	+28	+280	+26
Pacific Ocean	−4930	−6	+329	+5

added during the gravity reduction, the mass excesses of mountains or the deficiencies of oceans must be transferred to the roots or from the antiroots, respectively. This is accomplished in the most exact manner with the isostatic correction discussed.

To approach the problem in a different light, equate the Bouguer reduction with an isostatic reduction having an infinite level of compensation, or a free-air reduction with a zero compensation level (the surface of the geoid). Further, consider the meaning of zero gravity anomalies the world over. If all free-air anomalies were zero, all topography above the geoid would be without mass. For all Bouguer anomalies to be zero, mountains would indeed be excess piles of rubble on the geoid.

The only zero anomaly conditions with more than trivial meaning are those based on isostasy—in this case, they mean that the earth is in the equilibrium hypothesized for the reduction. Positive isostatic anomalies indicate undercompensation; such would be the case on a volcanic island so small it can be partially supported by the rigidity of the the crust. For instance, Oahu has isostatic anomalies of between +50 and +75 mgal; Cyprus, of 100 mgal. A negative anomaly means overcompensation, as in the case of a mountain wearing down by erosion faster than the compensating material disappears below it. The continents approach isostatic equilibrium quite closely: the United States is within a few milligals of being compensated except along the Pacific Coast, where the average is −20 mgal.

From the comparisons of Table 8–1, it appears that the free-air anomalies are close to those that result from isostasy with $T=40$ km. The mean value of free-air anomalies in a region are generally consistent with those stemming from an isostatic reduction. This consistency with the more adequate but rigorous isostatic reduction and its ease of application leads most practicing geodesists to champion the free-air reduction for use with the Stokes function.

SUMMARY

To conclude this discussion of the gravity method, let us review the steps in acquiring geodetic position at an astronomical station:

1. From reduced gravity measurements and the earth's angular velocity, determine the flattening of the reference ellipsoid with Eq. (8–11). The radius a must be obtained from other sources.

2. Obtain surface measurements of gravity on a world-wide basis. These should be based on the same initial assumption (Potsdam, etc.).

3. Reduce surface measurements to the geoid with, as an example, the free-air reduction shown in Eq. (8–27).

4. Compute the normal gravity, Eq. (8–13), for points on the ellipsoid corresponding to the surface stations.

5. With Eq. (8–36), compute the free-air anomalies.

6. With expressions similar to Eqs. (8–18) and (8–19) and global anomalies from step 5, compute gravimetric deflections of the vertical.

7. With these deflections applied to the astronomical coordinates, geodetic coordinates and azimuths on an ellipsoid centered at the earth's center of gravity are computed with Eqs. (8–21)–(8–23). Geodetic coordinates determined at any point in the world by this method will be consistent and will refer to the same ellipsoid, at least to a greater extent that those derived from any other method.

8. With Eq. (8–16), the distance between the ellipsoid and geoid (geoid height) may be computed. As leveling refers to the mean sea level equipotential surface, the geoid, the sum of the elevation and geoid height yields the departure of the terrain from the ellipsoid; therefrom, the size and the shape of the earth are determined.

REFERENCES

1. G. Bomford, *Geodesy*. Oxford: Clarendon Press, 2nd Ed. 1962.
2. W. A. Heiskanen and F. A. Vening Meinesz, *The Earth and Its Gravity Field*. New York: McGraw-Hill Book Co., 1958.

3. H. Orlin, Marine gravity surveying instruments and practice. *Proc. First Marine Geodesy Symposium*, Columbus, Ohio, September 28–30, 1966.

4. D. A. Rice, Deflections of the vertical from gravity anomalies, *Bull. Geod.* No. 25, 1952.

5. Urho A. Uotila, *Investigations on the Gravity Field and Shape of the Earth*. Publication of the Institute of Geodesy, Photogrammetry and Cartography of the Ohio State University, No. 10, Columbus, Ohio, 1960.

6. P. S. Zakatov, *A Course in Higher Geodesy*. Jerusalem: The Israel Program for Scientific Translations for the National Science Foundation, Washington, D.C., and the Department of Commerce, U.S.A., 1962 (translated from Russian).

Chapter 9 Leveling

The applications of direction, distance, and gravity measurements have been discussed, but to complete any introduction to geodesy, the concepts of elevation and height are needed. A point is only partially located on the earth by determining its latitude and longitude.

The elevation of a point is acquired by leveling. For geodetic quality, this is normally accomplished by differential levels. A telescope whose optical axis is parallel to the horizon plane is sighted fore and aft on graduated rods held over two points of interest. The difference of the rod readings yields the elevation between the two points. The intent, however, is not to explore the mechanics of this measurement. The interested reader is referred to standard texts on surveying for elaboration on this as well as the techniques of trigonometric and barometric leveling.

The practice of leveling is inherently linked to level surfaces, which are surfaces on which a marble does not roll freely. The purpose of this chapter is to look at the nature of these level surfaces that are the product of the earth's gravity field.

Just as horizontal geodetic coordinates must be referred to a geodetic datum, so must elevations. Geodetic height, h, is the distance between the point on the earth and the reference ellipsoid of the geodetic datum. This height is the sum of the distance between the geoid and the ellipsoid—geoid height, N—and the distance between the geoid and the point—the elevation, H. (This N should not be confused with the N used to signify the radius of curvature in the prime vertical.) Therefore, the reference surface or datum for elevations is the geoid or mean sea level.

It should be noted that the actual sea level surface may depart from the geoid at any given time. These departures are caused by the periodic gravitational effects of the sun and moon (tides), wind, changes in barometric pressure, differences in water densities, and glacial melting. For a land area, the sea level datum is defined as the mean sea level, determined from the tidal cycles averaged over at least a one-year period of observations. However, as some causes of these variations are not strictly periodic, all oceans are not on the same mean level. The mean level of the Pacific appears to be 72 cm higher than that of the Atlantic. Differential levels across the Isthmus of Panama show the Pacific to be about 20 cm higher than the Atlantic [3]. Lack of observations and of a fixed base of reference for sea level in the open ocean have been obstacles to formation of a global level datum.

The datum for elevations used in the United States is the Sea Level Datum of 1929 (SLD29). Such a datum represents the geoid and, as such, is an equipotential surface—a level surface with a constant potential. Each point whose elevation is referred to this level datum is on another level surface of constant potential. It is the relationship between these two surfaces that is of interest.

To investigate this relationship, consider the sea level datum to be a rotating sphere of radius R_1 whose potential U_1 is expressed by Eq. (8–8),

$$U_1 = \frac{GM}{R_1} + \frac{\omega^2 R_1{}^2 \cos^2 \theta}{2},$$

where θ is spherical latitude and ω is the rotational velocity. For a point on the equator, the potential is

$$U_1 = \frac{GM}{R_1} + \frac{\omega^2 R_1{}^2}{2}.$$

Now suppose there is a second surface of radius R_2 through a station of elevation $R_2 - R_1$ at the equator. Its potential is

$$U_2 = \frac{GM}{R_2} + \frac{\omega^2 R_2{}^2}{2}.$$

Subtracting this potential from that of the datum yields

$$U_1 - U_2 = \Delta U = GM\left(\frac{1}{R_1} - \frac{1}{R_2}\right) + \frac{\omega^2}{2}(R_1{}^2 - R_2{}^2)$$

$$= \frac{GM}{R_1 R_2}(R_2 - R_1) - \frac{\omega^2}{2}(R_2 - R_1)(R_2 + R_1).$$

Employing the approximations

$$R_1 R_2 = R_1{}^2 \qquad \text{and} \qquad R_1 + R_2 = 2R_1$$

produces

$$\Delta U = (R_2 - R_1)\left(\frac{GM}{R_1{}^2} - \omega^2 R_1\right).$$

Recall from Chapter 8 that

$$F = -\frac{GM}{R^2} = \text{gravitation},$$

$C_e = \omega^2 R =$ centrifugal force at the equator, and
$g_e = F + C_e =$ gravity at the equator,

then, apparently,

$$\Delta U = -(R_2 - R_1)\, g_e$$

and the distance between the surfaces at the equator is

$$(R_2 - R_1)_e = -\frac{\Delta U}{g_e}. \qquad (9\text{--}1)$$

The negative sign in Eq. (9–1) indicates that as R increases, U decreases, as explained in Chapter 8.

Running through the same procedure for the same two surfaces at the pole produces

$$\Delta U = \frac{GM}{R_1{}^2}\, (R_2 - R_1).$$

Recall that $C_p = 0$ (centrifugal force at the poles) and that $g_p = F = -GM/R^2$; then

$$\Delta U = -g_p(R_2 - R_1)$$

and the distance between the surfaces is

$$(R_2 - R_1)_p = -\frac{\Delta U}{g_p}. \qquad (9\text{--}2)$$

As these are assumed to be the same equipotential surfaces dealt with at the equator, ΔU is constant; therefore,

$$(R_2 - R_1)_p \neq (R_2 - R_1)_e.$$

As g_p is larger than g_e, $(R_2 - R_1)_p$ is smaller than $(R_2 - R_1)_e$. The conclusion drawn from this result is that the two level surfaces which are parallel at the equator approach one another toward the pole, i.e., the distance between these two level surfaces is not constant.

This convergence is illustrated by the International Gravity Formula, Eq. (8–14). At the equator, normal gravity is

$$\gamma_e = 978.049 \text{ gal;}$$

and at the pole,

$$\gamma_p = 983.221 \text{ gal.}$$

If the distance between the surfaces at the equator is

$$(R_2 - R_1)_e = 200 \text{ m} = 200 \times 10^2 \text{ cm}$$

and γ_e is substituted for g_e, Eq. (9–1) shows that the potential difference between the surfaces will be

$$\Delta U = - \; 1956.098 \text{ cm}^2/\text{sec}^2.$$

With γ_p and this constant value of ΔU in Eq. (9–2), the distance between the surfaces at the pole is found to be

$$(R_2 - R_1)_p = 198.948 \text{ m}.$$

These two surfaces are 1.052 meter closer at the poles than at the equator. Had the distance between the surfaces been 2000 m, at the poles they would have been closer by 10.52 m.

Although this phenomenon is not a major problem in most cases, it cannot be ignored in the theory.

The basic equation for this chapter is formed by writing Eq. (9–1) or (9–2) in general terms:

$$H = - \frac{\Delta U}{g}, \tag{9–3}$$

where H is the elevation of a point above or below a reference level surface, ΔU is the potential difference between this surface and that through the point, and g is gravity. Owing to changes of gravity on the earth's surface, it is readily apparent from Eq. (9–3) that the distance between surfaces will vary.

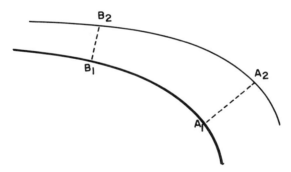

Figure 9–1. Converging level surfaces.

The result of leveling without gravity is shown in Fig. 9–1. The surface A_1B_1 is the reference surface; that through A_2 and B_2 is another level surface above A_1B_1. If differential levels were observed along A_1A_2 and then north from A_2 to B_2 along the level surface, the elevation of B_2 would be equal to A_1A_2. On the other hand, if the elevation of B_2 were determined by direct measurement from B_1, B_1B_2 would be

the verdict. Of course, B_1B_2 does not equal A_1A_2 and two different elevations for B_2 are available depending on the route taken.
In the same way, if leveling were accomplished from A_1 to A_2 to B_2 to
B_1 and closed back on A_1, a misclosure of $A_1A_2 - B_2B_1$ would be observed due to the converging surfaces. Elevation determinations which
are dependent on the route of leveling do not result in single values and
this leaves something to be desired.

The problem is how to obtain unique elevations independent of the
leveling route over which they were determined. Equation (9–3) and
gravity are the keys to the solution.

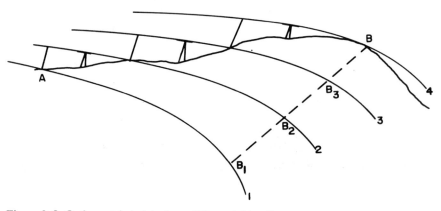

Figure 9–2. Orthometric height from differential leveling.

The following discussion is related to Fig. 9–2, which depicts a series
of differential level measurements from station A on the reference surface, or surface 1, up a hill to station B. The level surfaces through
intervening stations, or turning points, and station B are numbered 2,
3, and 4. Surface gravity at each point is signified by g_A, g_2, g_3, and g_B.
Also, ΔH_{A-2}, ΔH_{2-3}, and ΔH_{3-B} indicate the *observed* difference in elevation between surfaces; and $\Delta H'_{1-2}$, $\Delta H'_{2-3}$ and $\Delta H'_{3-B}$ indicate the
corresponding distances between these surfaces along the vertical under
B. The internal gravity existing at the intersection of this vertical and
these surfaces (points B_3, B_2, and B_1) is symbolized as g_1', g_2', and g_3'.

As the potential difference between surfaces is constant, Eq. (9–3)
may be written for these observations:

$$\Delta H_{A-2}\, g_A = -\,\Delta U_{1-2} = \Delta H'_{1-2}\, g'_1,$$
$$\Delta H_{2-3}\, g_2 = -\,\Delta U_{2-3} = \Delta H'_{2-3}\, g'_2,$$
$$\Delta H_{3-B}\, g_3 = -\,\Delta U_{3-4} = \Delta H'_{3-B}\, g'_3.$$

Upon summation, these equations yield the orthometric height, H_0, that is, the vertical distance from B to the reference surface:

$$H_0 = \sum \Delta H' = \left(\Delta H_{A-2} \frac{g_A}{g_1'} + \Delta H_{2-3} \frac{g_2}{g_2'} + \Delta H_{3-B} \frac{g_3}{g_3'} \right)$$

$$= \sum \left(\frac{\Delta Hg}{g'} \right). \tag{9-4}$$

This orthometric height is the desired quantity. It is the actual distance between the station and the reference level datum and is independent of the route of leveling, provided all gravity values are known.

However, gravity is the hitch in the determination of orthometric heights. Surface gravity can be obtained economically, but the internal gravity along the vertical of the end station can only be hypothesized.

A value for the internal gravity may be computed by using techniques of gravity reduction similar to those discussed in Chapter 8. For instance, g_B is observed on the surface of the earth and the value of gravity at the intersection of the vertical and the reference surface is sought; that is, the g_1' existing at B_1. First, the mass between B and B_1 is removed with the Bouguer correction (densities: $p = 2.67$ and $p_m = 5.576$);

$$\delta_B' = -0.1108 \text{ mgal/meter.}$$

Then the point is moved in free air to B_1 with the free-air correction;
$$\delta_f' = 0.3086 \text{ mgal/meter.}$$

Thus far, we have been developing a Bouguer reduction for land. However, our object is not gravity on a clean geoid as required for computation of a gravity anomaly, but rather actual gravity with the external masses in place. Therefore, the mass initially removed must be returned with a second Bouguer correction;

$$\delta_B' = -0.1108 \text{ mgal/meter.}$$

Combining these corrections, we compute the gravity value at B_1:

$$g_1' = g_B +0.0870 \text{ mgal/meter.}$$

(The terrain correction may also be included in this sequence, depending on extremes of area terrain.) This is the Prey reduction and offers a method of computing internal gravity.

This correction may be computed and applied to each leveling increment in the summation of Eq. (9-4). A reasonable labor-saving approach, however, is to employ a mean value for the gravity existing between B and B_1, that is,

$$\bar{g}' = \frac{g_B + g_1'}{2}.$$

With this mean, Eq. (9–4) is written

$$H_0 = \frac{\Sigma(\Delta Hg)}{\bar{g}'}. \tag{9–5}$$

There are other approaches to the gravity question of orthometric heights. References [1] and [2] contain excellent material on this subject.

The orthometric height achieved with Eq. (9–4) satisfies the requirement for a unique height. It represents the actual height of the station above the reference surface. As just explained, however, this height is not obtained without employing assumption; it is only obtainable in theory. A conceptual disadvantage is that orthometric heights along a level surface are not constant.

An alternate approach that offers a solution for this last criticism is to substitute a constant value for gravity in place of $\bar{g}'$ in the denominator of Eq. (9–5). When the normal gravity at latitude $45°$ is employed, the so-called dynamic height is achieved:

$$H_D = \frac{\Sigma(\Delta Hg)}{\gamma_{45°}}. \tag{9–6}$$

With Eq. (9–3), recall that $\Sigma(\Delta Hg) = -\Delta U =$ constant potential difference between surfaces and is, therefore, independent of the leveling route used to connect these surfaces. Obviously, H_D is a constant value on a level surface.

Though expressed in linear units, this H_D has no geometric significance in the real world of the geoid and the terrain. This objection may be overcome by using the geopotential number of the level surface rather than the dynamic height. This geopotential number is simply

$$\Delta U = \Sigma(\Delta Hg). \tag{9–7}$$

ΔU is expressed in geopotential units (gpu) where 1 gpu = 1000 gal-meters. As $\Delta U \doteq g\,H \doteq 0.98H$, it is apparent that the geopotential number assigned to a surface is nearly its height above sea level in meters.

There is another height system that lies between the extremes offered by orthometric and dynamic heights. If the earth's gravity is assumed normal, normal gravity could be substituted in Eq. (9–5) to produce normal heights, H_N. Normal gravity values are easily obtained with the International Gravity Formula or other such conventions. These normal values are independent of route. Like orthometric heights, they are not constant along a level surface. Because the earth's gravity and potential are not normal, these values have little direct meaning in the leveling

objectives discussed. Normal heights find significance in modern gravimetric theories. Such applications were introduced by Molodenskii and
are explained in [2] and [4].

SUMMARY

1. As are other geodetic coordinates, elevations are referred to a
datum. Such a datum for a given land mass is normally established as
the observed mean sea level. A datum thus defined for land areas
separated by oceans may not be the same level surface.

2. This datum is one equipotential surface resulting from the earth's
mass and rotation. Any two such surfaces are closer together at the
poles than they are at the equator. The potential difference between
surfaces is constant; the distance between them is not. The elevation of
a point determined by leveling along these surfaces is dependent on
whether the levels were run to the north or south.

3. Orthometric height is the distance from the geoid to the point. It is
independent of the leveling route but requires values for gravity at the
surface leveling stations and along the vertical under the terminal
station. The use of a computed or mean value for the latter interior
gravity limits the theoretical cleanliness of this determination. Note,
however, that if the mean value of gravity used in Eq. (9–5) fails to
represent the true condition along the vertical by 50 mgal for a 2000 m
elevation, the resulting error in the computed orthometric height will
be only 10 cm.

4. Just as observed directions must account for deflections of the
vertical to achieve geometric consistency, so must level observations
give gravity its just due.

REFERENCES

1. C. F. Baeschlin, *Lehrbuch der Geodasie*. Zürich, Orell-Füssli, Verlag, 1948.
2. W. A. Heiskanen and H. Moritz, *Physical Geodesy*. W. H. Freeman and Company, 1967.
3. I. Hela and E. Lisitzin, A world mean sea level and marine geodesy. In *Proc. First Marine Geodesy Symposium*, Columbus, Ohio, September 28–30, 1966.
4. R. A. Molodenskii, *et al*, *Methods for Study of the External Gravitational Field and Figure of the Earth*. Jerusalem: The Israel Program For Scientific Translation with the National Science Foundation, Washington, D.C., and the Department of Commerce, U.S.A., 1962 (translated from Russian).

Chapter 10   Satellites

In the field of geodesy, satellites (artificial or otherwise) have made global surveys a realistic objective. For the first time, oceans may be effectively spanned to connect continents, and geodetic positions may be determined directly on a global datum with origin at the earth's center of gravity (CG).

Our natural satellite, the moon, permits the principle of the eclipse to become as a key to geodesy. The moon obscures a star (occultation) or the sun (solar eclipse). The observed phenomena are functions of the observer's position on earth; therefore, if stations record observation times along the eclipse path, their positions may be related.

These methods have serious observational limitations. Solar eclipses are infrequent and when they occur provide solutions only for stations on the path of the shadow cast by the moon. Star occultations may be observed more frequently; however, the irregular and undetermined profile of the moon suggests that this method has only limited accuracy. (This uncertainty is fading with current lunar research.) Both methods are generally weather sensitive. The solar eclipse expeditions of 1945, 1947, 1948, 1952, 1954, and 1955 achieved poor to fair results [6]. In the 1945 expedition sponsored by the Air Force Cambridge Research Laboratories, all but two of the 16 stations were overcast. The relative accuracy estimated for the solar eclipse solution is 1/75,000 [9].

Because of the foregoing limitations, artificial satellites, with their greater operational flexibility, will be the subject explored in this chapter.

Artificial satellites are observed electronically or optically to measure range, range rate, or direction from the observer to the satellite. The mathematical analysis of such observations produce relative positions among observing sensors, sensor position on a global datum, or define the satellite orbit. Each of these processes will be explored.

Logically, the first thing to explore is the orbit of satellites, how they are defined, and how they may be determined. Next, measurements of direction, range, or range rate to the satellite are developed to determine geodetic position. These measurements are performed under two conditions—satellite orbit known, and satellite orbit not required. The systems that make these measurements and methods of reducing observations for input in the geodetic solution are discussed. For each system, some operational project results are given. Lastly, the types of geodetic satellites are described. This chapter is concluded with a table summarizing the characteristics of systems and various operational modes.

SATELLITE ORBIT

An understanding of the path traveled by the satellite is interesting and necessary background for subsequent discussions of this geodetic tool. The basic equations will be cited with minimum derivation and, although these forms could be used, more sophisticated mathematics suitable for electronic computers are employed in practice.

An orbiting satellite travels in an elliptical path with the earth's center at one of the foci. This ellipse (as all others discussed) may be defined by its semi-major axis, a and the square of its eccentricity, e^2. If $e = 0$, the ellipse is a circle; the satellite's position is simply a function of time and polar coordinates. When the path is a true ellipse, however, this calculation becomes difficult. Not only does the distance from the center of the earth to the satellite vary, but also its angular velocity is not constant.

The elements of the orbit plane required to approach this dilemma are pictured in Fig. 10–1. The angle v is termed the *true anomaly* and the angle E, the *eccentric anomaly*. There is a fictitious anomaly, M, called the *mean anomaly*. M corresponds to a v of uniform rate and is expressed simply as a function of the *mean angular rate*, n, and time; that is,

$$M = n\,(t - T);\qquad(10\text{–}1)$$

where

$$n = \frac{2\pi}{P},\qquad(10\text{–}2)$$

and P is the period of the satellite orbit; T is the time of perigee passage (closest approach to orbit foci); and t is the time of satellite position after T.

The intermediary M is related to the geometric E by Kepler's equation,

$$M = E - e \sin E.\qquad(10\text{–}3)$$

If the satellite position is to be determined as a function of time, this equation is solved by iteration. Mueller [13] outlines the solution as follows:

First compute M for a time t with Eq. (10–1), and then compute an approximate E_1 with

$$E_1 = M + e \sin M + 1/2\,e^2 \sin 2M.$$

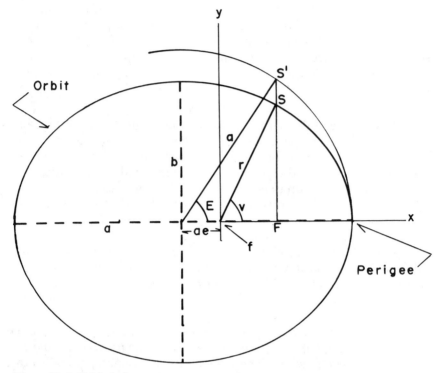

Figure 10–1. Orbital plane.

Substituting this value in Eq. (10–3), compute M_1.

Now with $\Delta M = M_1 - M$, determine a correction for E_1 with

$$\Delta E = \frac{\Delta M}{1 - e \cos E};$$

then compute

$$E_2 = E_1 + \Delta E.$$

With E_2, repeat this sequence until $\Delta M = 0$ (or nearly so).

    With E known, r and v are now sought. By referring to the auxiliary x and y coordinate axis in Fig. 10–1, note that

$$x = r \cos v = a \cos E - ae = a(\cos E - e), \qquad (10\text{–}4)$$

and

$$y = r \sin v = b \sin E = a(1 - e^2)^{1/2} \sin E . \qquad (10\text{–}5)$$

Equation (10–5) is derived by noting the ratio $b/a = SF/S'F$, and recalling that $b = a(1 - e^2)^{1/2}$. From these expressions, the following may be written easily:

$$r = (x^2 + y^2)^{1/2} = a(1 - e \cos E), \tag{10-6}$$

and

$$\tan v = \frac{y}{x} = \frac{(1 - e^2)^{1/2} \sin E}{\cos E - e}. \tag{10-7}$$

Prior to closing the discussion of the plane elliptical orbit, two more interesting relations should be noted. If a satellite were placed motionless above the earth's surface, it would move in only one direction—toward the earth—due to gravitation. If, however, an appropriate velocity component perpendicular to this path existed, the satellite would orbit as discussed. The following equation expresses this necessary velocity component, s', tangent to the elliptical path:

$$s'^2 = GM \left(\frac{2}{r} - \frac{1}{a} \right), \tag{10-8}$$

where GM is the product of the gravitation constant and earth's mass as used in Chapter 8, and r and a are as illustrated in Fig. 10–1. Equation (10–8) is known as the vis-viva equation.

The second item of interest is stated by Kepler's Second Law; that is, the area velocity (the area swept by r in a unit of time) is constant. This condition is written

$$A' = \tfrac{1}{2} \bar{r} s' = \text{constant},$$

where $\bar{r}$ is the perpendicular distance from the foci to the satellite track projected tangent to the orbit path ($\bar{r} \neq r$). If Eq. (10–8) is substituted for s' and the satellite is considered at the end of the minor axis (where $r = a$ and $\bar{r} = b$); then

$$A' = \frac{b}{2} \left(\frac{GM}{a} \right)^{1/2} = \text{constant}. \tag{10-9}$$

Dividing the area of the ellipse ($ab\pi$) by the area velocity, the orbital period is produced:

$$P = \frac{ab\pi}{A'} = \frac{2\pi a^{3/2}}{(GM)^{1/2}}.$$

Substituting this expression in Eq. (10–2) produces Kepler's Third Law,

$$n^2 a^3 = GM = \text{constant}, \tag{10-10}$$

that is, the period squared is proportional to the semi-major axis cubed.

The satellite motion and position in the plane of the ellipse have been defined. The position of the satellite in the *celestial coordinate system* now deserves some attention.

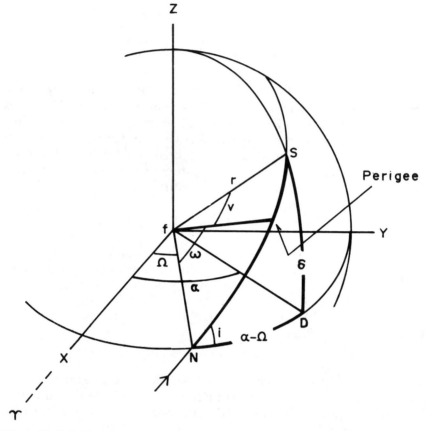

Figure 10–2. Orbital elements.

Equations (10–6) and (10–7) yield polar coordinates that must be transformed to celestial coordinates of right ascension, α, and declination, δ. Figure 10–2 illustrates this problem. The X-axis is toward the vernal equinox, Υ; the Z-axis coincides with the mean rotation axis of the earth; and the Y-axis is perpendicular to both. The line formed by the intersection of the orbit and the equatorial plane is the *line of nodes*. As portrayed, N is the *ascending node*. The line from the foci, f, through perigee is the *line of apsides*. Additionally, Ω is the right ascension of the ascending node, ω is the argument of perigee, and i is the angle of inclination between the orbital and equatorial planes.

The a, e, P, and T specify the plane elliptical orbit, and the angles Ω, ω, and i position this plane on the celestial sphere. These seven parameters are termed the *Keplerian elements* of orbit.

The right ascension and declination of the satellite may be determined by solving the right spherical triangle NDS,

$$\sin \delta = \sin (\omega + v) \sin i, \qquad (10\text{–}11)$$

and

$$\tan (\alpha - \Omega) = \cos i \tan (\omega + v). \qquad (10\text{–}12)$$

If the satellite were orbiting a uniform spherical earth in a vacuum, the foregoing equations would suffice. Unfortunately, these conditions do not exist and the Keplerian elements are not constant, and the orbit, therefore, is *perturbed.*

The earth's flattening and atmospheric drag are two causes of perturbation for near-earth orbits. The fact that the earth is flattened at the poles causes the orbital plane to rotate about the earth's axis in the opposite direction of satellite travel—a change in Ω; and furthermore, for the semi-major orbit axis to rotate in the plane of the orbit, a change in ω. On the other hand, drag causes the major orbit axis to shorten and produces a reduction in eccentricity.

To obtain expressions for perturbations, recall that the satellite orbit is a response to the potential of the earth's gravitation (see Chapter 8). The potential for an earth assumed to be an ellipsoid may be expressed in a series as

$$V = \frac{GM}{r} \left[1 - \sum_{n=2}^{\infty} \mathcal{J}_n \left(\frac{a_e}{r} \right)^n P_n(\cos \theta) \right];$$

where $P_n(\cos \theta)$ are the Legendre polynomials of degree n (zonal spherical harmonics), θ is the co-latitude of the satellite, $\mathcal{J}_n$ are the constant coefficients, and a_e is the semi-major axis of the earth's ellipsoid. In this chapter, the subscript e is used to prevent confusion with the orbit semi-major axis, a. (Note: these expressions appear in the literature in several forms using different notation.)

Similarly, the rate of change of the Keplerian elements, Ω and ω, due to the gravitation of this ellipsoid may be expressed as a function of the orbital elements and constant $\mathcal{J}_n$:

$$\frac{d\Omega}{dt}, \frac{d\omega}{dt} = f(n, a, e, i, \mathcal{J}_n).$$

Specifically, the almost linear effects (secular) may be expressed by considering only $\mathcal{J}_2$:

$$\frac{d\Omega}{dt} = -n \left(\frac{a_e}{a(1 - e^2)} \right)^2 3/2 \, \mathcal{J}_2 \cos i; \qquad (10\text{–}13a)$$

$$\frac{d\omega}{dt} = +n \left(\frac{a_e}{a(1 - e^2)} \right)^2 3/2 \, \mathcal{J}_2 \, (1 + \cos^2 i - 3/2 \sin^2 i). \qquad (10\text{–}13b)$$

If the J_2 coefficient is known, the rate of change in Ω and ω may be computed. On the other hand, if $d\Omega$ and $d\omega$ are observed, a set of equations may be written and solved for the value of J_2. This proposition will be discussed later.

The other Keplerian elements can be expressed as functions of the disturbing influence of drag, attraction of the moon and sun (lunisolar effects), radiation pressure, tides, and relativity. A more extensive series with longitude-dependent terms may be written to account for the fact that the earth's gravity field is not simply the result of an ellipsoid of uniform mass distribution.

In passing, note that the effects on the orbit of flattening and drag and irregular mass distribution of the earth diminish with an increase of orbit height.

A detailed theory of orbit perturbations is beyond the scope of this book. The interested reader is referred to [3], [11], and [13].

In practice, the problem of computing satellite position with the varying elements of orbit can be solved by assuming the variable ellipse to be tangent to the actual orbit at the satellite point. Such a changing ellipse is said to osculate (to kiss) with the true orbit. If all disturbing forces were removed, the satellite would travel the path defined by the ellipse at the moment of osculation.

The Astrophysical Observatory of the Smithsonian Institute, Space Track, and other agencies publish values for the Keplerian elements and the rate at which they are changing.

To recapitulate the subject of orbit prediction, assume that the elements of orbit at time of perigee passage, T, are known—Ω_T, ω_T, i_T, a_T, e_T, and P. Also given are the rates of change of the major elements, $d\Omega/dt$ and $d\omega/dt$. Furthermore, with the assumption that these rates are linear, the elements at the desired time, t, may be computed:

$$\Omega_t = \Omega_T - \frac{d\Omega}{dt}\,(t-T)\;;$$

$$\omega_t = \omega_T - \frac{d\omega}{dt}\,(t-T)\,.$$

With Eq. (10–2) and P, n is computed, and with Eq. (10–1), the mean anomaly, M_T, is computed. At this point, the process can be refined by computing M_t:

$$M_t = M_T + \frac{d\omega}{dt}\,(t-T),$$

recalling that v, E, and, therefore, M, are referred to perigee which is defined by ω. With Eq. (10–3), compute the eccentric anomaly, E_t. With Eq. (10–7), the true anomaly, v_t, is computed. Likewise, Eq.

(10–6) yields the distance, r, from the center of the earth to the satellite. With v_t, ω_t, Ω_t, and i_T, solution of Eqs. (10–11) and (10–12) results in the right ascension, α_t, and declination, δ_t, at t.

If they are required for purposes of acquisition, the azimuth and the elevation angle from an observer to the satellite (*look angles*) can be computed by the equations of Chapter 6. With the same logic employed in deriving Eqs. (6–1)–(6–3), the geodetic rectangular coordinates of the satellite are written:

$$U_S = r \cos \delta \cos (\alpha - H),$$
$$V_S = r \cos \delta \sin (\alpha - H),$$
$$W_S = r \sin \delta,$$

where r has replaced the N, h, and e^2 terms; δ is substituted for ϕ; and $(\alpha - H)$ for λ. The H is the angle in units of sidereal time between the vernal equinox and the Greenwich meridian at t (Greenwich Sidereal Time).

The observer's rectangular geodetic coordinates—U_M, V_M, W_M—are computed with Eqs. (6–1)–(6–3). The ΔU, ΔV, and ΔW obtained with coordinates of satellite and observer, and the geodetic coordinates of the observer, are then applied to Eqs. (6–17) and (6–18) to yield the geodetic azimuth, α, and the vertical angle, v, to the satellite. Specifically,

$$\tan \alpha = \frac{\Delta U \sin \lambda - \Delta V \cos \lambda}{\Delta U \cos \lambda \sin \phi + \Delta V \sin \lambda \sin \phi - \Delta W \cos \phi}, \qquad (10\text{--}14)$$

$$\sin v = \frac{\Delta U \cos \lambda \cos \phi + \Delta V \sin \lambda \cos \phi + \Delta W \sin \phi}{(\Delta U^2 + \Delta V^2 + \Delta W^2)^{1/2}}. \qquad (10\text{--}15)$$

(Note the different meanings symbolized by α and v.)

Now that the factors that define an orbit have been presented, consider a general manner of determining orbit elements [3]. Assume a satellite's position and velocity have been observed at t (without pondering how). This means that r, dr, and s' are known. Equation (10–8) will yield the value of a:

$$a = \frac{1}{\left(\dfrac{2}{r} - \dfrac{s'^2}{GM} \right)}.$$

From Eq. (10–6), we write

$$e \cos E = 1 - \frac{r}{a}. \qquad (10\text{--}16)$$

Because there are two unknowns, E and e, another equation is required. By differentiating Eq. (10–16),

$$dr = a e \sin E \, dE,$$

and by differentiating Eq. (10–3),

$$dM = n = (1 - e \cos E)\, dE.$$

Multiplying by a and referring to Eqs. (10–6) and (10–10), dE may be expressed

$$dE = \frac{na}{r} = \left(\frac{GM}{a}\right)^{1/2} \frac{1}{r}.$$

Substituting this expression for dE into that for dr produces

$$e \sin E = \frac{r\, dr}{(GMa)^{1/2}}. \qquad (10\text{–}17)$$

Dividing this equation by Eq. (10–16) yields

$$\tan E = \left(\frac{a}{GM}\right)^{1/2} \left(\frac{r\, dr}{a - r}\right)$$

which may be solved for E. Either Eq. (10–16) or (10–17) may be used to solve for e. From Eq. (10–10), the mean motion, n, may be determined:

$$n = \left(\frac{GM}{a^3}\right)^{1/2}.$$

With E and n now known, Eq. (10–1) is equated with Eq. (10–3) and the resulting expression is solved for T, the time of perigee passage:

$$T = \frac{e \sin E - E}{n} + t.$$

With Eq. (10–4), the true anomaly is computed:

$$\cos v = \frac{a\,(\cos E - e)}{r}.$$

We have the polar elements of the orbit plane—a, e, n and T, as well as v and r at t. Now, Ω, ω and i must be determined. An easy way to approach this problem is to refer back to Eqs. (10–11) and (10–12). If sin $(\omega + v)$ is written as $\sin v \cos \omega + \cos v \sin \omega$ in Eq. (10–11), then

$$\sin \delta = \sin v \cos \omega \sin i + \cos v \sin \omega \sin i.$$

It is apparent there are two unknowns in this equation, $\cos \omega \sin i$ and $\sin \omega \sin i$. The satellite position, α and δ, are known and the true anomaly, v, was just computed. With two observations, two simultaneous equations in variables δ_1, δ_2, α_1, and α_2 may be written and solved for the unknowns. Then,

$$\tan \omega = \frac{\sin \omega \sin i}{\cos \omega \sin i}.$$

With ω thus determined and substituted in Eq. (10–11), the inclination is computed,

$$\sin i = \frac{\sin \delta}{\sin v \cos \omega + \cos v \sin \omega}.$$

With these values of i and ω, Eq. (10–12) may be used to compute $\alpha - \Omega$, from which Ω may be extracted.

The foregoing illustrates an approach to orbit determination. Satellite position could have been determined by simultaneous observations from sets of two or more cameras. Theoretically, of course, the position of observing cameras should be on a datum whose center coincides with the earth's CG.

In practice, the mathematical form used to determine satellite orbit parameters depends upon the type of observation—direction, range, range rate, or a combination [3]. The elements would normally be overdetermined (more observations than unknowns) and the end result would be produced by a least squares adjustment (see Chapter 11).

Before leaving the discussion of satellite orbits, consider the geodetic information provided solely by knowledge of the orbit. Through successive observations of a satellite's travel, the changes in its orbital elements are detected. As indicated in Eq. (10–13), these changes may be expressed with the same constant terms, the J_n terms, used in expressions for the geopotential. Obviously, with such constant terms, a mathematical expression for the earth's potential and gravity may be written. The adequacy of this representation depends on the degree to which a series is carried and how well the effects of nongravitational forces have been removed.

If you will recall, $d\Omega$ is a function of the earth's flattening and may be expressed in terms of the J_n values. It follows that an expression containing flattening and the J_n terms may be written. A mean value for flattening, f, may be approximated with

$$f = \frac{q}{2}(1-f) + \frac{3}{2}J_2 + \frac{5}{8}J_4;$$

where q is the ratio of centrifugal and gravitational acceleration at the equator.

In the current literature the foregoing items are often termed dynamic applications of satellite geodesy.

GEODETIC POSITION FROM KNOWN ORBIT

As explained in the last section, coordinates of an orbiting satellite are referred to a system whose origin coincides with the earth's center of gravity (CG). By observing a satellite of known position, the observer may be located in this CG-centered system, Again the specific form of the solution depends on the type of observation. In this section, geodetic position resulting from direction, range, and range-rate observations are discussed.

A single camera makes *direction observations* of an illuminated satellite (flashing light or chopped trace from a reflecting surface) by photographing it against a background of stars. With the time of exposure known, the right ascension and declination of the stars photographed can be calculated. This information makes it possible to compute the observed celestial coordinates of the satellite. (The photographic plate computations are discussed later in this chapter.) With a defined orbit, the actual position of the satellite is also known. The observed coordinates will not equal the orbit coordinates (except under very special circumstances). This coordinate difference is the basis for positioning the observer.

Figure 10–3 portrays two rectangular coordinate systems—one originated at the earth's CG (the X, Y, Z system shown in Fig. 10–2) and

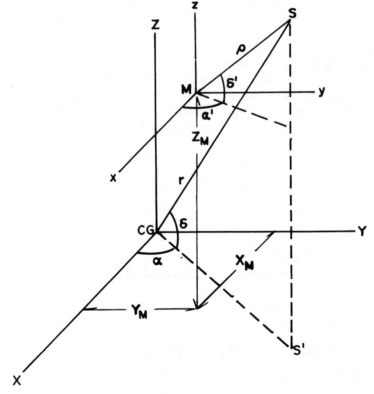

Figure 10–3. Celestial and topocentric coordinate systems with direction observations.

the other with origin at the observer (x, y, z topocentric system). The coordinate axes of these systems are parallel. The X-axis is toward ♈ and the Z-axis coincides with the axis of rotation of the earth.

Furthermore, α and δ are known satellite celestial coordinates and are those that would be viewed by an observer at the earth's center; α' and δ' are the celestial coordinates actually observed (topocentric); ρ is the range from the observer to the satellite; r is the range from the satellite to the earth's center; and X_M, Y_M, and Z_M are the rectangular coordinates of the observer in the celestial coordinate system at the moment of observation.

In the XYZ system,

$$\cos \alpha = \frac{X}{r \cos \delta},$$

$$\sin \alpha = \frac{Y}{r \cos \delta},$$

$$\sin \delta = \frac{Z}{r}.$$

Thus, the satellite coordinates can be expressed as

$$X = r \cos \delta \cos \alpha, \tag{10–18a}$$

$$Y = r \cos \delta \sin \alpha, \tag{10–18b}$$

$$Z = r \sin \delta. \tag{10–18c}$$

Similarly, in the xyz system, the topocentric system,

$$\sin \delta' = \frac{z}{\rho}, \quad \text{and} \quad \rho = \frac{z}{\sin \delta'}.$$

Therefore, the satellite coordinates in this system are

$$x = \rho \cos \delta' \cos \alpha' = z \cot \delta' \cos \alpha', \tag{10–19a}$$

$$y = \rho \cos \delta' \sin \alpha' = z \cot \delta' \sin \alpha', \tag{10–19b}$$

and, of course,

$$z = \rho \sin \delta'. \tag{10–19c}$$

(Note that ρ is not observed.)

Now, as both coordinate systems are parallel,

$$X = X_M + x, \tag{10–20a}$$

$$Y = Y_M + y, \tag{10–20b}$$

$$Z = Z_M + z. \tag{10–20c}$$

Obviously,

$$z = Z - Z_M,$$

and, substituting $\mathcal{Z}$ from Eqs. (10–18),

$$z = r \sin \delta - \mathcal{Z}_M.$$

Furthermore, substituting this value for z in Eqs. (10–19a, b), then substituting these x and y values on the right side of Eqs. (10–20a, b) and the X and Y expressions of Eqs. (10–18a, b) on the left yields

$$r \cos \delta \cos \alpha = X_M + (r \sin \delta - \mathcal{Z}_M) \cot \delta' \cos \alpha', \qquad (10\text{–}21\text{a})$$

$$r \cos \delta \sin \alpha = Y_M + (r \sin \delta - \mathcal{Z}_M) \cot \delta' \sin \alpha'. \qquad (10\text{–}21\text{b})$$

By rearranging so that all terms with r are on the left; then multiplying the first equation by $\sin \alpha'$, the second by $\cos \alpha'$, adding, and then applying the trigonometric identity for the sine of the difference of angles; one obtains

$$r \cos \delta \sin (\alpha' - \alpha) = X_M \sin \alpha' - Y_M \cos \alpha'. \qquad (10\text{–}22)$$

Again, by multiplying Eq. (10–21a) by $\cos \alpha'$, and Eq. (10–21b) by $\sin \alpha'$, adding, recalling that $\sin^2 \alpha' - \cos^2 \alpha' = 1$, and using the identity for the cosine of the difference of angles, we obtain

$$r[\cos (\alpha' - \alpha) \cos \delta - \sin \delta \cot \delta'] = X_M \cos \alpha' + Y_M \sin \alpha' - \mathcal{Z}_M \cot \delta'. \qquad (10\text{–}23)$$

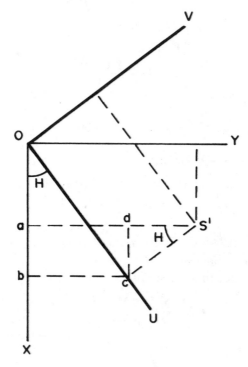

Figure 10–4. Coordinate transformation in equatorial plane.

Unfortunately, the earth rotates in the celestial XYZ system and, therefore, X_M, Y_M, and Z_M will vary with time. To overcome this sticky problem, the XYZ system will be expressed in terms of the fixed geodetic rectangular system, UVW. This transformation is accomplished simply by rotating the XYZ system around the Z-axis (which is assumed to coincide with the W-axis) by the hour angle H between the vernal equinox and the Greenwich meridian. (The UVW system is described in Chapter 6 and in this case is assumed to have its origin at the earth's CG.) H corresponds to the time of the satellite flash. Figure 10–4 illustrates this relationship and the necessary elements for transformation.

$$X = Oa = Ob - ab = Ob - dc = Oc \cos H - S'c \sin H,$$
$$= U \cos H - V \sin H; \tag{10–24a}$$

$$Y = S'a = S'd + ad = S'd + bc = S'c \cos H + oc \sin H,$$
$$= V \cos H + U \sin H; \tag{10–24b}$$

$$Z = W. \tag{10–24c}$$

Substituting these equations in Eqs. (10–22) and (10–23), and again applying the trigonometric identities for the sine and cosine of angular differences, results in

$$r \cos \delta \sin (\alpha' - \alpha) = U_M \sin (\alpha' - H) - V_M \cos (\alpha' - H); \tag{10–25}$$

$$r [\cos (\alpha' - \alpha) \cos \delta - \sin \delta \cot \delta'] = U_M \cos (\alpha' - H)$$
$$+ V_M \sin (\alpha' - H) - W_M \cot \delta'. \tag{10–26}$$

(Note that the foregoing equations are of the same form as Eqs. (10–22) and (10–23) with UVW substituted for XYZ and $(\alpha' - H)$ substituted for α'; henceforth, such substitutions will constitute this transformation.)

Equations (10–25) and (10–26) are written for every observation. Two observations result in four equations which may be solved for U_M, V_M, and W_M. The quantities α' and δ' are observed at a time corresponding to H; r, α, and δ are computed from the known orbit.

Now consider the proposition when *range* rather than direction is observed to a satellite of known orbit. The SECOR system to be discussed later is in this category.

Again, Eqs. (10–6), (10–11), and (10–12) would be used to compute r, δ, and α for the satellite at the time of ranging. Likewise, Eqs. (10–18) yield its XYZ position. With the identical philosophy but opposite approach employed in deriving Eq. (10–24), the following equations may be written to transform satellite position to the CG-centered geodetic system:

$$U = X \cos H + Y \sin H, \tag{10–27a}$$

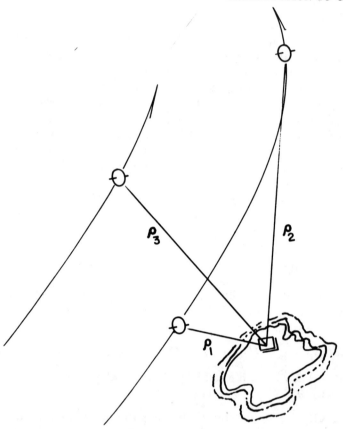

Figure 10–5. Range observations on three satellite points.

$$V = -X \sin H + Y \cos H, \qquad (10\text{--}27b)$$
$$W = Z. \qquad (10\text{--}27c)$$

Figure 10–5 depicts range observations, ρ, on three satellite points. These three ranges can be written in terms of satellite and observer coordinates:

$$\rho_1{}^2 = (U_1 - U_M)^2 + (V_1 - V_M)^2 + (W_1 - W_M)^2, \qquad (10\text{--}28a)$$
$$\rho_2{}^2 = (U_2 - U_M)^2 + (V_2 - V_M)^2 + (W_2 - W_M)^2, \qquad (10\text{--}28b)$$

and

$$\rho_3{}^2 = (U_3 - U_M)^2 + (V_3 - V_M)^2 + (W_3 - W_M)^2. \qquad (10\text{--}28c)$$

Equations (10–28) present three equations in three unknowns, the observer's coordinates—U_M, V_M, and W_M. Because these are second-degree equations, however, their simultaneous solution for these unknowns is not simple.

One method of solving these equations is to differentiate range, ρ, in terms of the differential of the unknown coordinates:

$$dp_1 = \frac{(U_M{}^0 - U_1)\, dU}{\rho_1} + \frac{(V_M{}^0 - V_1)\, dV}{\rho_1} + \frac{(W_M{}^0 - W_1)\, dW}{\rho_1}, \quad (10\text{-}29a)$$

$$dp_2 = \frac{(U_M{}^0 - U_2)\, dU}{\rho_2} + \frac{(V_M{}^0 - V_2)\, dV}{\rho_2} + \frac{(W_M{}^0 - W_2)\, dW}{\rho_2}, \quad (10\text{-}29b)$$

$$dp_3 = \frac{(U_M{}^0 - U_3)\, dU}{\rho_3} + \frac{(V_M{}^0 - V_3)\, dV}{\rho_3} + \frac{(W_M{}^0 - W_3)\, dW}{\rho_3}. \quad (10\text{-}29c)$$

Approximate values for station coordinates, $U_M{}^0$, $V_M{}^0$, and $W_M{}^0$, are used to compute approximate ranges, $\rho_1{}^0$, $\rho_2{}^0$, and $\rho_3{}^0$. In Eqs. (10–29), the $d\rho$ represents the difference between the observed and the computed ρ^0. These equations are formed and solved simultaneously for dU, dV, and dW. These differentials are the corrections to the assumed station coordinates, that is,

$$U_M = U_M{}^0 + dU,$$
$$V_M = V_M{}^0 + dV,$$
$$W_M = W_M{}^0 + dW.$$

The concept employed with Eqs. (10–29) is applicable to this case where the satellite position is known and more than the minimum of three required satellite positions are available. The benefits and limitations of this method are fully developed in the discussion of least squares in Chapter 11.

From a geometric point of view, a single pass containing all three satellite points would produce a very weak solution. The short arc of the observed orbit would have so little curvature that it would approach a straight line. Fixed rays from this line may be drawn to intersect at a point off this line; however, this point may be rotated around the line without changing the length of these intersecting rays. For this reason at least two satellite passes are required to provide a realistically unique solution.

Range rate. In the U.S. Navy's Transit Network (TRANET) system, the change in radial range from an observer to a passing satellite is determined by observing the doppler shift in a transmitted signal from a satellite. Classically, doppler shifts are compared to the whistle of a passing train; the whistle has a high pitch on approach and a low pitch on departure. This is expressed by

$$F = F_T\left(1 + \frac{\Delta\rho}{C}\right),$$

and

$$\Delta\rho = \frac{d\rho}{dt} = \frac{C(F - F_T)}{F_T}; \quad (10\text{-}30)$$

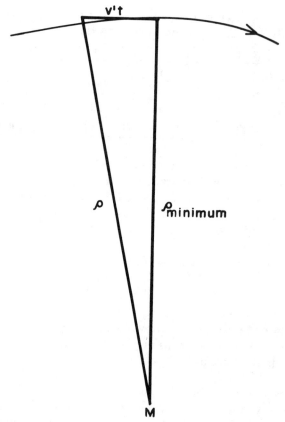

Figure 10–6. Minimum range with doppler observations.

where $\Delta\rho$ is the time rate of change of range, F_T is the transmitted frequency, F is the received frequency, and C is the propagation velocity. With F observed, and F_T and C known, $\Delta\rho$ is computed.

Furthermore, Fig. 10–6, shows that

$$\rho^2 = \rho^2{}_m + (v't)^2, \tag{10–31}$$

where, v' is the tangential velocity relative to the observer, ρ_m is the range at closest approach, and t is the time from the point of closest approach. In Eq. (10–30), observe that $F = F_T$ when ρ equals ρ_m. Differentiating Eq. (10–31) relative to ρ and t produces

$$\rho\frac{d\rho}{dt} = v'^2 t$$

which, upon squaring and substituting ρ from Eq. (10–31), results in

$$\frac{\rho_m{}^2}{v'^4} + \frac{t^2}{v'^2} = \frac{t^2}{\varDelta\rho}. \tag{10–32}$$

Theoretically, the time of closest approach is observed, that is when $F = F_T$; therefore, t is assumed known as well as $\varDelta\rho$ from Eq. (10–30), and ρ_m and v' are the unknowns. With Eq. (10–32) written for two observations at t_1 and t_2, the constants $1/v'^2$ and $\rho_m{}^2/v'^4$ may be solved and then ρ_m computed.

This treatment is primarily illustrative rather than practical. F_T and, in turn, the time of closest approach are not known precisely. Multiple observations are solved by the methods of least squares for uncertainties in F_T, the time of closest approach, and C, as well as the unknown, ρ_m.

With ρ_m determined for three passes of satellites of known orbits, Eqs. (10–29) may be solved for the geodetic coordinates of the observer.

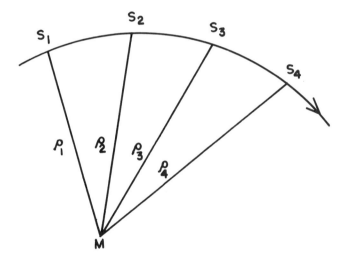

Figure 10–7. Range rates.

Another approach presented by [12] sheds light on the geometric interpretation of the doppler count. Figure 10–7 pictures a satellite positioned along its orbit at equal intervals—S_1, S_2, S_3, S_4, $\cdots S_n$ at t_1, t_2, t_3, t_4, $\cdots t_n$. Suppose, moreover, that the satellite emits signals at these times and that these signals are received on the ground at $t_1 + \varDelta t_1$, $t_2 + \varDelta t_2$, $t_3 + \varDelta t_3$, $t_4 + \varDelta t_4$, $\cdots t_n + \varDelta t_n$. The $\varDelta t$ is the time required for the signal to travel from the satellite to the observer at M.

We now add another frequency to the previous development, a

constant reference frequency, F_K, in the receiver. The station counts the number of beat cycles $\mathcal{N}_R$, formed by the difference in reference and received frequencies between time signals 1 and 2, and so on. This count is symbolized, for interval S_1 to S_2

$$\mathcal{N}_R = \int_{t_1+\Delta t_1}^{t_2+\Delta t_2} (F_K - F)\, dt,$$

which, on integration, is written

$$\mathcal{N}_R = F_K(t_2 - t_1) + F_K(\Delta t_2 - \Delta t_1) - \int_{t_1+\Delta t_1}^{t_2+\Delta t_2} F\, dt.$$

The remaining integral,

$$\int_{t_1+\Delta t_1}^{t_2+\Delta t_2} F\, dt,$$

is the total number of cycles received between $t_1 + \Delta t_1$ and $t_2 + \Delta t_2$ and, obviously, these must equal those transmitted between t_1 and t_2. Therefore,

$$\int_{t_1+\Delta t_1}^{t_2+\Delta t_2} F\, dt = \int_{t_1}^{t_2} F_T = F_T\,(t_2 - t_1);$$

and, by substituting this in the expression for $\mathcal{N}_R$,

$$\mathcal{N}_R = F_K(t_2 - t_1) + F_K(\Delta t_2 - \Delta t_1) - F_T(t_2 - t_1),$$
$$\mathcal{N}_R = (F_K - F_T)\,(t_2 - t_1) + F_K(\Delta t_2 - \Delta t_1). \qquad (10\text{–}33)$$

The term $(\Delta t_2 - \Delta t_1)$ is the difference in time to travel paths ρ_2 and ρ_1. Clearly,

$$\rho_2 - \rho_1 = C(\Delta t_2 - \Delta t_1);$$

therefore, with Eq. (10–33), the distance difference is written

$$\rho_2 - \rho_1 = \frac{C}{F_K} [\mathcal{N}_R - (F_K - F_T)\,(t_2 - t_1)]. \qquad (10\text{–}34)$$

All elements on the right are observed or are known constants:
$(t_2 - t_1)$ is the time interval between signal transmission,
F_K and F_T are constant frequencies,
C is the propagation velocity, and
$\mathcal{N}_R$ is the observed cycle count.

The distance differences thus determined are used to position the observer in a manner similar to that presented in Chapter 7 in the explanation of how distance differences define hyperbolas. Recall that range is expressed as

$$\rho_1{}^2 = (U_1 - U_M)^2 + (V_1 + V_M)^2 + (W_1 - W_M)^2$$

where U_1, V_1 and W_1 are the known satellite coordinates at t_1. With the same expression formed for a subsequent satellite point, S_2, a range difference equation is available, that is,

$$\rho_2{}^2 - \rho_1{}^2 = (\rho_2 - \rho_1)(\rho_2 + \rho_1)$$
$$= 2(U_1 - U_2)\, U_M + 2(V_1 - V_2)\, V_M + 2(W_1 - W_2)\, W_M$$
$$+ U_2{}^2 + V_2{}^2 + W_2{}^2 - U_1{}^2 - V_1{}^2 - W_1{}^2 = Q_{1,2}. \qquad (10\text{--}35)$$

Unfortunately, $\rho_2 + \rho_1$ is not observed. This problem is handled by adding the square of the observation to both sides of the equation:

$$(\rho_2 - \rho_1)^2 + (\rho_2 - \rho_1)(\rho_2 + \rho_1) = Q_{1,2} + (\rho_2 - \rho_1)^2;$$
$$(\rho_2 - \rho_1)\, 2\rho_2 = Q_{1,2} + (\rho_2 - \rho_1)^2.$$

The remaining unknown term, ρ_2, may be eliminated. First, write Eq. (10–35) for the next time interval, $t_3 - t_2$, and then subtract $(\rho_3 - \rho_2)^2$ from both sides of the result:

$$(\rho_3 - \rho_2)\, 2\rho_2 = Q_{2,3} - (\rho_3 - \rho_2)^2.$$

Then, multiplying the first equation by $-(\rho_3 - \rho_2)$, the second by $(\rho_2 - \rho_1)$, and adding, eliminates ρ_2:

$$(\rho_2 - \rho_1)\, Q_{2,3} - (\rho_3 - \rho_2)\, Q_{1,2}$$
$$= + (\rho_3 - \rho_2)^2 (\rho_2 - \rho_1) + (\rho_2 - \rho_1)^2 (\rho_3 - \rho_2).$$

By inserting the Q terms, this equation may be written in the following general form:

$$2[U_i\,(\rho_k - \rho_j) + U_j\,(\rho_i - \rho_k) + U_k\,(\rho_j - \rho_i)]\, U_M$$
$$+ 2[V_i(\rho_k - \rho_j) + V_j(\rho_i - \rho_k) + V_k(\rho_j - \rho_i)]\, V_M$$
$$+ 2[W_i(\rho_k - \rho_j) + W_j(\rho_i - \rho_k) + W_k(\rho_j - \rho_i)]\, W_M$$
$$= U_i{}^2(\rho_k - \rho_j) + U_j{}^2(\rho_i - \rho_k) + W_k{}^2(\rho_j - \rho_i)$$
$$+ V_i{}^2(\rho_k - \rho_j) + V_j{}^2(\rho_i - \rho_k) + W_k{}^2(\rho_j - \rho_i)$$
$$+ W_i{}^2(\rho_k - \rho_j) + W_j{}^2(\rho_i - \rho_k) + W_k{}^2(\rho_j - \rho_i)$$
$$+ (\rho_k - \rho_j)(\rho_i - \rho_k)(\rho_j - \rho_i). \qquad (10\text{--}36)$$

This equation has three unknowns, observed range differences and known satellite coordinates for three points, S_i, S_j, S_k; where

$$i = 1, 2, 3 \cdots n, \qquad j > i, \, k > j.$$

Three equations must be written for solution with Eq. (10–36). Points 1, 2 and 3 form one equation, points 2, 3 and 4 form another, and points 3, 4 and 5 form the third equation.

For similar reasons explained in conjunction with ranging techniques, a strong single solution cannot be gained for one pass.

Summary. All of the preceding applications for satellites of known orbits have one thing in common: they yield only the rectangular co-ordinates of the observer in a system whose origin coincides with the earth's center of gravity. The difference between these coordinates and those of the observer on the local datum is the displacement of the origin of the local system from the CG. In this way, a local or regional datum can be translated to a common origin—a major step in the geodetic problem.

It is emphasized, however, that these applications tell nothing about the size (a) or shape (f or e^2) of the best-fitting reference ellipsoid. With Eqs. (6–4)–(6–6), these CG referenced rectangular coordinates may be transformed to latitude, longitude, and geodetic height with any chosen a and f.

GEODETIC POSITION BY SIMULTANEOUS OBSERVATION, ORBIT NOT REQUIRED

The distance over which a surveyor can observe directions is limited by the earth's curvature, and slightly enhanced by atmospheric refraction. For example, from a 100-foot tower, a survey target mounted on another 100-foot tower can be seen no farther away than 26 miles (with no objects intervening). On the other hand, a satellite at a 500-mile height could be viewed by an observer looking 10° above the horizon 1300 miles away. This is exactly how satellites are used in the *intervisible mode*, i.e., the satellite is merely an elevated survey target that can be sighted by several observers simultaneously.

Much of the thought presented in the preceding section pertaining to known orbits still applies. Here, instead of knowing the satellite position in the CG-centered coordinate frame, the satellite is positioned in the coordinate system from which it is observed. A sensor at an unknown location observes simultaneously with sensors of known location and can be located in their system with the same mathematics used previously. Essentially, only the problem of determining the satellite's position has changed. This concept is readily illustrated in the following discussion with *direction observations*.

Equation (10–20) shows the relation among satellite coordinates, observer coordinates, and observed coordinates. Now, however, the satellite coordinates are unknown. The terms on the right are known or observed. There is one other difference: although the coordinate axes still are parallel to those of the inertial system previously treated, the system does not necessarily have its origin at the earth's CG. The origin is now the center of the geodetic datum upon which the observers are located. To prevent confusion, these coordinates will be indicated with an overbar, that is, $\bar{X}$, $\bar{Y}$, $\bar{Z}$, $\bar{U}$, $\bar{V}$, and $\bar{W}$. There is no change in the so-called observed coordinates, x, y, and z.

With the above exceptions, the procedures used to obtain Eqs. (10–25) and (10–26) are used. In Eqs. (10–20), the observed x and y expressions of Eqs. (10–19) and $z = \bar{Z} - \bar{Z}_M$, in turn, are substituted. The resulting $\bar{X}$ equation is multiplied by $\sin \alpha'$; the $\bar{Y}$ by $-\cos \alpha'$ and added; the $\bar{X}$ is then multiplied by $\cos \alpha'$ and $\bar{Y}$ by $\sin \alpha'$ and again added. In the two equations produced by these additions, the relationships of Eqs. (10–24) are substituted. The preceding operations result in the following expressions:

$$\bar{U} \sin (\alpha' - H) - \bar{V} \cos (\alpha' - H) = \bar{U}_M \sin (\alpha' - H) - \bar{V}_M \cos (\alpha' - H); \tag{10–37a}$$

$$\bar{U} \cos (\alpha' - H) - \bar{V} \sin (\alpha' - H) - \bar{W} \cot \delta' = \bar{U}_M \cos (\alpha' - H)$$
$$+ \bar{V}_M \sin (\alpha' - H) - \bar{W}_M \cot \delta'. \tag{10–37b}$$

Recapitulating: α', δ', H are observed; $\bar{U}_M$, $\bar{V}_M$, $\bar{W}_M$ are known; and $\bar{U}$, $\bar{V}$, $\bar{W}$ are to be determined.

With two or more cameras of known position observing the satellite simultaneously, four or more equations result and may be solved for the three unknowns. The satellite is positioned by *intersecting* rays from the ground.

If a camera of unknown position observes the satellite at the same instant as the known cameras and does this on two or more occasions, its position also may be determined with Eqs. (10–37). In this reverse case, $\bar{U}_M$, $\bar{V}_M$, and $\bar{W}_M$ are the unknowns. As each observation yields two equations, at least two such occurrences are required. The unknown observer is located by *resection*.

This method of *intersection–resection* necessitates simultaneous observation by three cameras of at least two satellite points. It specifically computes the satellite position.

There is another approach presented by [18]. This method solves the problem with intersecting planes, uses simultaneous observation by two cameras, and does not specifically compute the satellite position. Use of the two-camera observation has operational advantages and, because the satellite position is not computed, eases the computation burden.

The necessary mathematics may be derived by substituting Eqs. (10–19) in Eqs. (10–20) and writing the resulting expressions for simultaneous observation from cameras A and B.

$$\bar{X} = \bar{X}_A + \rho_A \cos \delta'_A \cos \alpha'_A, \qquad \bar{X} = \bar{X}_B + \rho_B \cos \delta'_B \cos \alpha'_B;$$
$$\bar{Y} = \bar{Y}_A + \rho_A \cos \delta'_A \sin \alpha'_A, \qquad \bar{Y} = \bar{Y}_B + \rho_B \cos \delta'_B \sin \alpha'_B;$$
$$\bar{Z} = \bar{Z}_A + \rho_A \sin \delta'_A, \qquad \bar{Z} = \bar{Z}_B + \rho_B \sin \delta'_B.$$

Obviously, these six equations may be combined (with care) to eliminate satellite coordinates $\bar{X}$, $\bar{Y}$, and $\bar{Z}$, and range terms ρ_A and ρ_B. Then, substituting $\bar{U}$, $\bar{V}$, $\bar{W}$ and $(\alpha' - H)$ for $\bar{X}$, $\bar{Y}$, $\bar{Z}$ and α', respectively, we may write:

$$(\bar{U}_A - \bar{U}_B) [\tan \delta'_B \sin (\alpha'_A - H) - \tan \delta'_A \sin (\alpha'_B - H)]$$
$$- (\bar{V}_A - \bar{V}_B) [\tan \delta'_B \cos (\alpha'_A - H) - \tan \delta'_A \cos (\alpha'_B - H)]$$
$$+ (\bar{W}_A - \bar{W}_B) \sin (\alpha'_B - \alpha'_A) = 0. \tag{10–38}$$

This equation may be written for every two-camera set of simultaneous observations. It may be considered the equation of a plane passing through both cameras and the satellite. For a second simultaneous observation set, another plane is defined which intersects the first plane along the line connecting the cameras. Clearly, however, the equation for this line is not sufficient to solve for the coordinates of the unknown camera, symbolized as camera B. Additional observations between these two stations do not solve the problem. An observation set with an additional known camera yields a plane which the line will intersect at the unknown camera.

Minimum requirements for positioning the unknown station with Eq. (10–38) are three sets of simultaneous observations with the unknown camera and one set with two known cameras.

In passing, note that two sets of simultaneous observations between two stations are adequate for an azimuth determination. The unknown station lies along the line defined by the intersection of the two planes. With A assumed known, and with one coordinate arbitrarily selected for B, $\bar{W}_B$ for instance, two sets of Eq. (10–38) may be solved for values of $\bar{U}_B$ and $\bar{V}_B$ on this line that correspond to $\bar{W}_B$. With the resulting $\Delta \bar{U} = \bar{U}_B - \bar{U}_A$, $\Delta \bar{V}$, $\Delta \bar{W}$, and Eqs. (10–14) and (10–15), the azimuth and vertical angle between A and B may be computed. Though rather trivial with satellites, this feature could be employed to advantage with an aircraft equipped with a flashing light and flying between otherwise nonintervisible ground stations.

Now the application of ranging to *intervisible* satellite observations is explored. First, the satellite's position must be determined. Equations (10–29)—the differential solution—could be written for range observation from three ground stations of known position to the satellite of unknown position. By simultaneous solution, these equations could be solved for corrections to coordinates assumed for the satellite. In the case of known orbit, the unknowns need be estimated only once regardless of the number of observations used in the solution. In the present case, however, the satellite coordinates must be estimated for each set of three observations.

An approach presented by [14] alleviates some of the work. The range equations, Eqs. (10–28), are of the second degree in both observer and satellite coordinates. These equations may be simplified with an auxiliary coordinate system, U' V' W'. This system has its origin at one station (station 1), has its V' axis through station 2, and all three stations are on the $U'V'$ plane. With this system, Eqs. (10–28) are written (subscript M changes to S for satellite positions and numerical subscripts now represent stations):

$$\rho_1{}^2 = U'_S{}^2 + V'_S{}^2 + W'_S{}^2, \qquad\qquad (10\text{--}39a)$$
$$\rho_2{}^2 = U'_S{}^2 + (V'_2 - V'_S)^2 + W'_S{}^2, \qquad (10\text{--}39b)$$
$$\rho_3{}^2 = (U'_3 - U'_S)^2 + (V'_3 - V'_S)^2 + W'_S{}^2. \qquad (10\text{--}39c)$$

Whereas Eqs. (10–28) could not be solved easily, Eqs. (10–39) may be solved in a straightforward manner for the satellite positions. When three such satellite positions have been observed simultaneously by an unknown ground station, its position may be determined with Eqs. (10–29). This position in the auxiliary coordinate system may be transformed back to the U, V, and W system by applying the translation and rotation employed in establishing the U', V', and W' system originally.

As noted previously, two separate satellite passes are required to maintain geometric integrity.

Lastly, look at possible applications of *range rate* with simultaneous observations. This approach, called "translocation," determines locations of two nearby stations, as in the case with the orbit known. Then, however, one position solution is subtracted from the other to yield a position difference or relative position—ΔU, ΔV, and ΔW. As the error in the satellite position is assumed equal for both solutions, and would thereby cancel in differencing, the relative set of coordinates is essentially free from orbital uncertainties.

Specifically, Eq. (10–36) is written for sets of observations made from stations A and B on a satellite at five identical time intervals. The same initial position for the satellite is assumed for calculations from both stations. The coordinates for A and B are computed and differenced to obtain relative positions. Symbolically, for the U coordinate,

$$U_A + E_U + E\,\frac{dU_i}{dt} = f\left(U_i + E_U,\, U_i + \frac{dU_i}{dt} + E\,\frac{dU_i}{dt}\right),$$

$$U_B + E_U + E\,\frac{dU_i}{dt} = f\left(U_i + E_U,\, U_i + \frac{dU_i}{dt} + E\,\frac{dU_i}{dt}\right),$$

where E_U is the error in assumed initial satellite coordinates, dU_i/dt is the change in satellite position and $E(dU_i/dt)$ is the error in this assumption. Now,

$$U_A - U_B + E_U - E_U + E\,\frac{dU_i}{dt} - E\,\frac{dU_i}{dt} = U_A - U_B.$$

$E(dU_i/dt)$ is considered small, which is reasonable over the small orbit interval observed. The validity of the assumption that orbital error cancels depends on the distance between the observing stations.

In summary, the common feature of these techniques requiring no knowledge of the satellite orbit is that more than one station must simultaneously observe satellites at more than one time—with directions, three observers and two satellite points; with range, four observers and three satellite points; with range rate, two observers and five satellite points. Coordinates thus determined are in the same system as those of the known observers.

Between these extremes of known orbits and one observer, and unknown orbits with multiple observers, there is an approach known as the *short-arc technique*. Although multiple observers are required, it is not necessary for the unknown station to observe simultaneously with known observers.

Multiple stations of known position observe a passing satellite and thereby fix its position at a given time and, possibly, determine its motion. From these coordinates, satellite position is extended until it is observable by the unknown station. In this way, a distant observer is located in a known coordinate system. The basic premise of this concept is that uncertainties of satellite motion for the short distance from known locations contribute a smaller error to the solution than that of a solution based on pure orbital calculations. (With present techniques, it is estimated that satellite position may be calculated to about 300 feet accuracy with orbital elements.) Of course, the greater the distance between known and unknown observers, the greater the error. Arc methods are usually recommended for a distance of less than one revolution.

Other types of equations may be formed to solve the problem both with and without a known orbit. As in the case of ranging and a known orbit, Eqs. (10–29), mathematical reductions based on differentials may be used when observations are compared with those computed from assumed station coordinates. Equations are written to express this comparison as a function of corrections to the assumed stations coordinates. Likewise, processes may be formed to adjust not only station coordinates but satellite position and motion as well [13].

In passing, note that combined observations greatly simplify the problem presented. As an example, with range and direction observed, Eqs. (10–18), (10–19) (with the ρ term), (10–20), and (10–24) are combined to yield:

$$U = R \cos \delta \cos (\alpha - H) - \rho \cos \delta' \cos (\alpha' - H),$$
$$V = R \cos \delta \sin (\alpha - H) - \rho \cos \delta' \sin (\alpha' - H),$$
$$W = R \sin \delta - \rho \sin \delta'.$$

One observation of α', δ', and ρ on a satellite of known orbit produces three equations with which to solve three unknowns. Combined observations of range and range rate also produce simplified and stronger solutions. Such combined observing systems are in development. The range and direction system (laser) is discussed briefly in the next section.

OBSERVATION TOOLS

Solutions for the observer's position with satellite observations of direction, range, and range rate have been explored. Next, the systems that make these observations will be investigated. Only those systems considered mobile are presented.

Unlike theodolite observations of terrestrial directions, all of the satellite techniques to be discussed are completely independent of the direction of the vertical.

Direction Instruments. Precise directions are observed with cameras, which provide photographs of illuminated satellites against a background of stars. There are basically two types of satellites used. One has a flashing light—an active satellite, and one merely reflects sunlight or other external light source—a passive satellite.

Procedures for obtaining useful data vary. With the active satellite, the shutter is opened for the period during which the satellite is in view and its flash images are photographed as points of light. On the other hand, a passive satellite would produce a streak of light for any extended exposure and, therefore, the shutter must be opened and closed rapidly to reduce the streak to points of light—the light streak is chopped.

To obtain the desired directions to the satellite images, they are referred to the photographed stellar background. Like the passive satellite, stars produce streaks of light during extended exposure and these traces are chopped. The star calibration sequence consists of exposures of varying duration to obtain the best measurable image for stars of different brightness. Exposure duration during this sequence may range from 2 to 0.1 seconds. Time is recorded for each shutter action. Such a calibration sequence is carried out before and after each satellite observation to insure sufficient redundancy to minimize random errors. As the area of the heavens being observed is known, stars are readily identified and their coordinates computed for the time of the exposure.

The essential elements of this procedure are as follows. After the stars are identified, their mean coordinates for a given year are found in a star catalog—for instance, the Boss catalog, epoch 1950, listed in the Chapter 5 reference list. These coordinates must be corrected to mean coordinates for the year of observation and then to the time of observation by methods fully presented in the Explanatory Supplement to the *American Ephemeris and Nautical Almanac.* Coordinates thus determined are termed the "true position."

Next, the right ascension and declination, α_0 and δ_0, of the point assumed to be the center of the photographic plate are determined. This may be accomplished by using the celestial coordinates of a star near the center of the plate, by determining the center of the field of stars used for calibration (the centroid method), or by measuring the geometric center of the plate at the same time the stars are measured and determining the center coordinates by linear interpolation.

The next step is to compute the rectangular coordinates of the stars in a plane tangent to the celestial sphere at the projection of the optical axis of the camera. These coordinates cannot be measured directly on the camera plate. This problem is depicted in Fig. 10–8. C is the optical center of the camera and is assumed to coincide with the center of the XYZ system—a perfectly valid assumption considering that the celestial sphere is of infinite radius. The coordinates of the star projected on the

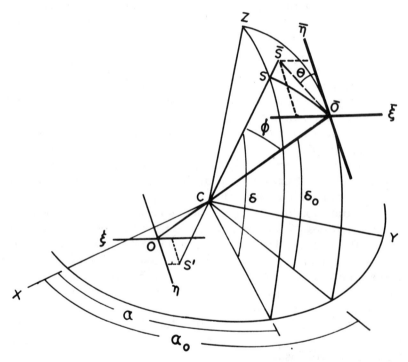

Figure 10–8. Standard coordinates.

tangent plane are $\bar{\eta}$ and $\bar{\xi}$ with $\bar{\eta}$ positive toward the celestial pole and $\bar{\xi}$ positive in the direction of increasing right ascension and perpendicular to $\bar{\eta}$ and the optical axis, $C\bar{O}$. Note that line $\bar{S}\bar{O}$ is arc $S\bar{O}$ projected and the angle θ in the plane equals the spherical angle $Z\bar{O}S$. In the tangent plane,

$$\bar{\eta} = \bar{O}\bar{S} \cos \theta = C\bar{O} \tan \phi \cos \theta = \tan \phi \cos \theta \qquad (10\text{–}40)$$

with $C\bar{O} = 1$; likewise,

$$\bar{\xi} = \tan \phi \sin \theta. \qquad (10\text{–}41)$$

Referring to the spherical triangle $\overline{Z}\overline{O}S$, note that arc $\overline{Z}S = 90° - \delta$, arc $\overline{Z}\overline{O} = 90° - \delta_0$, and angle $S\overline{Z}\overline{O} = \alpha - \alpha_0$ From the laws of sines and cosines for a spherical triangle,

$$\sin \phi \sin \theta = \cos \delta \sin (\alpha - \alpha_0), \tag{10-42}$$

$$\cos \phi = \sin \delta \sin \delta_0 + \cos \delta \cos \delta_0 \cos (\alpha - \alpha_0), \tag{10-43}$$

$$\cos \theta = \frac{\sin \delta - \cos \phi \sin \delta_0}{\sin \phi \cos \delta_0}.$$

Rewriting Eq. (10-42),

$$\sin \phi = \frac{\cos \delta \sin (\alpha - \alpha_0)}{\sin \theta},$$

then multiplying it by the value of $\cos \theta$ shown above and substituting Eq. (10-42) in the denominator will yield.

$$\cos \theta \sin \phi = \frac{\sin \delta - \cos \phi \sin \delta_0}{\cos \delta_0}.$$

In order to remove the unknown, $\cos \phi$, from the right side, Eq. (10-43) is substituted, which, upon some elementary simplification, produces

$$\cos \theta \sin \phi = \sin \delta \cos \delta_0 - \cos \delta \sin \delta_0 \cos (\alpha - \alpha_0). \tag{10-44}$$

Dividing Eq. (10-44) by Eq. (10-43) and using the expression obtained in Eq. (10-40), the relation for $\bar{\eta}$ is obtained:

$$\bar{\eta} = \frac{\cos \delta_0 - \cot \delta \sin \delta_0 \cos (\alpha - \alpha_0)}{\sin \delta_0 + \cot \delta \cos \delta_0 \cos (\alpha - \alpha_0)}. \tag{10-45}$$

To obtain $\bar{\xi}$, divide Eq. (10-42) by Eq. (10-43), that is,

$$\bar{\xi} = \frac{\cot \delta \sin (\alpha - \alpha_0)}{\sin \delta_0 + \cot \delta \cos \delta_0 \cos (\alpha - \alpha_0)}. \tag{10-46}$$

In Fig. 10-8, CO is the focal length; O is the center of the camera plate that is parallel to the tangent plane. On the camera plate, the η axis is parallel to the $\bar{\eta}$ axis but is pointed in the opposite direction. The same is true for the ξ and $\bar{\xi}$ axes. Angle $\overline{S}C\overline{O}$ equals angle $S'CO$, etc. Because all angular relationships are equal, we have similar prisms. As in the case for the tangent plane, Eqs. (10-45) and (10-46) will express the camera plate η, ξ coordinates of the star S in units of the focal length if $CO = 1$; remember,

$$\eta = CO \tan \phi \cos \theta, \quad \text{and} \quad \xi = CO \tan \phi \sin \theta.$$

These coordinates are referred to as "standard coordinates," and are computed for each star used in the plate calibration, as well as for the point assumed to be the center of the plate.

Next, the x and y coordinates measured on the plate must be related to the standard coordinates just computed. If measurements were error-free and observations were made in a uniform atmosphere from a nonrotating earth, the x and y plate coordinates would equal the η and ξ standard coordinates. To compensate for refraction difference for stars at different declinations, aberration caused by earth rotation, and measurement and observation errors, the following relations are written:

$$\xi - x = \xi a + \eta b + c \tag{10-47a}$$

and

$$\eta - y = \xi d + \eta e + f. \tag{10-47b}$$

By computing standard coordinates and measuring plate coordinates for three stars, six equations may be written and solved for the plate constants a, b, c, d, e, and f. From this process stems the requirement for a minimum of three stars for plate orientation; however, at least ten stars are desirable to achieve redundancy and reduce errors[1].

Next, the x and y coordinates of the satellite images are measured on the plate. With the plate constants just determined, and by using Eqs. (10-47), the standard coordinates of the satellite images, ξ_s and η_s, are computed. Furthermore,

$$\bar{\xi}_s = \frac{\xi_s}{f'} \quad \text{and} \quad \bar{\eta}_s = \frac{\eta_s}{f'};$$

that is, dividing the determined coordinates by the focal length, f', yields coordinates compatible with the system in the tangent plane.

The last step requires the reverse of the relationships defined by Eqs. (10-45) and (10-46), namely, α_s and δ_s as functions of $\bar{\eta}_s$ and $\bar{\xi}_s$. Multiplying Eq. (10-45) by its denominator and combining terms in δ and α on the left ($\delta = \delta_s$, $\alpha = \alpha_s$), and dividing the numerator and denominator by $\cos \delta_0$ yields

$$\cot \delta_s \cos (\alpha_s - \alpha_0) = \frac{1 - \bar{\eta}_s \tan \delta_0}{\bar{\eta}_s + \tan \delta_0}. \tag{10-48}$$

In the same manner, Eq. (10-46) may be juggled and written

$$\cot \delta_s \sin (\alpha_s - \alpha_0) = \frac{\bar{\xi}_s \sec \delta_0}{\bar{\eta}_s + \tan \delta_0},$$

which, when divided by Eq. (10-48), results in

$$\tan (\alpha_s - \alpha_0) = \frac{\bar{\xi}_s \sec \delta_0}{1 - \bar{\eta}_s \tan \delta_0}. \tag{10-49}$$

Obviously, this equation is solved for α_s and then Eq. (10-48) is approached for δ_s.

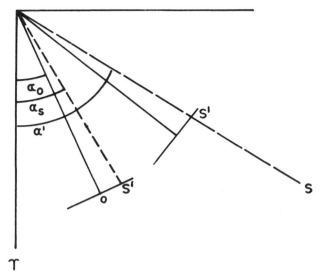

Figure 10–9. Direction and earth's rotation.

Before employing these satellite coordinates, they must be adjusted to correspond with those directions observed at the instant of flash, which will not agree with the instant of plate calibration. Recall that camera direction is defined from star images at a specific time. Since these images are normally recorded before or after satellite flashes, the camera axis will be pointing in a different direction at the instant of flash. In Fig. 10–9, a star makes an image O on the center of the camera plate at time t_0, and this is used to compute α_0 and δ_0 for the center of the standard coordinates. At the time t_1, satellite image S' is photographed. As developed, Eqs. (10–45) through (10–49) would produce α_s, which does not equal the required observed topocentric α' of Eqs. (10–25) and (10–26).

From this discussion, it is apparent that an adjustment must be made which may be stated as

$$\alpha' = \alpha_s + (\text{correction}) \, (t_1 - t_0).$$

This correction may be simply the earth's rotation rate.

For nonintervisible observations, directions on the celestial sphere must correspond to those at the instant of flash or chop; furthermore, for intervisible observations, directions from all participating cameras should be adjusted to a common time. There are several ways of effecting this adjustment.

Additionally, α' and δ' must be adjusted for differential refraction and aberration. As explained in conjunction with Eqs. (10–47), stellar refraction and aberration are accounted for in the transformation between x, y and ξ, η coordinates. Therefore, computed satellite directions carry the effect of the entire atmosphere and aberration at the sidereal rate. Because the light from the satellite does not travel as far as that from the stars, and because the velocity of the satellite either adds to or subtracts from the sidereal rate, a slight correction must be applied to account for this difference in refraction and aberration.

The α' and δ' resulting from these processes may be used in Eqs. (10–25) and (10–26) or in Eq. (10–38) for appropriate solution of camera position.

The procedures presented generally follow the concepts of classical astronomy. For a streamlined technique made possible with today's electronic computers, see [5].

To obtain the observations discussed, four elements are necessary: a well-calibrated camera with a low vibration shutter, a clock of constant rate, a radio to receive time signals, and a device to record shutter action and time.

The two ballistic cameras commonly employed—the Wild BC-4 with a 300 mm focal length and the 1000 mm PC–1000—are shown in

Figure 10–10. Wild BC–4 ballistic camera (Wild Heerbrugg Instruments, Inc.).

Figure 10–11. PC–1000 ballistic camera (official U.S. Air Force photograph).

Figs. 10–10 and 10–11. Observations with the PC–1000 are estimated to result in directions accurate to ± one second of arc, and with the BC–4 to ± two seconds of arc [4]. These estimates include the effects from all error sources.

Both cameras have internal shutters that chop star trails for camera orientation. If a passive satellite is being observed, a shutter system of higher rate is used. Such shutters may be externally mounted to avoid camera vibration. Internal shutters that will satisfy both requirements are being developed and tested.

A constant time base is normally provided by a crystal oscillator, which is generally stable to one part in 10^8. Such reference time code generators are synchronized with received time signals from WWV, WWVH, or stations broadcasting time and frequency standards.

Shutter action, continuous time standard, and received time signals are combined by means of a recording device. A 2- or 4-channel magnetic tape recorder is an example of such a device.

Before reviewing the accomplishments of these systems, we should mention laser development by the National Aeronautics and Space Administration and the U.S. Air Force Cambridge Research Laboratories. By directing a laser light pulse from the ground to a satellite-mounted reflector, direction as well as range may be ascertained. The pulse illuminates the satellite and the time of transit may be used for range computation. A major advantage of the laser system is that the complex instrumentation is kept on the ground. A major difficulty with this approach is the aiming of the laser signal at the satellite and vice versa [15]. A $\pm$2-meter ranging accuracy is estimated [4].

Mobile cameras have been used extensively in the intervisible mode of operation. Europe and the North American continent have been tied through a net of BC–4 cameras operated by the U.S. Coast and Geodetic Survey. The passive Echo satellites were observed. A joint effort between the U.S. Coast and Geodetic Survey and the Army Map

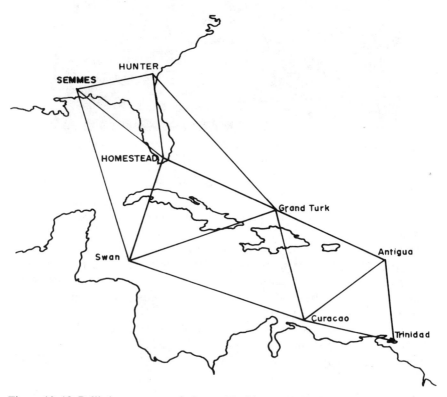

Figure 10–12. Ballistic camera geodetic net, Florida to Trinidad.

Service including more than 40 stations will circle the globe. Through observations of the high-altitude passive satellite PAGEOS (Passive Geodetic Earth Orbiting Satellite), these stations will be tied together to connect areas never before related.

Typical of projects using active optical satellites in the intervisible mode was the tie of Trinidad and other islands to the coast of Florida. Figure 10–12 indicates the eight stations occupied by USAF PC–1000 cameras. The cameras with known positions were located at Semmes, Alabama; Hunter Air Force Base, Georgia; and Homestead Air Force Base, Florida. A relative accuracy approaching one part in 500,000 was achieved in the reduction of the ANNA (Army–Navy–NASA–Air Force) and GEOS A (Geodetic Earth Orbiting Satellite, series A) observations by the U.S. Air Force Aeronautical Chart and Information Center.

In a subsequent project using the same cameras equipped with external chopping shutters, Bermuda was connected to stations in Georgia and Maryland by observing passive satellites. A relative accuracy of near one part per million was reported [10].

Electronic Ranging Instruments. The system employing this principle for the geodetic goal is known as SECOR, an acronym for Sequential Collation of Range. This system determines range by measuring phase shifts [14].

The ground stations transmit to the satellite on a frequency of 420.9 MHz. This carrier wave is phase-modulated in four different measurement frequencies—585.533 KHz, 36.596 KHz, 2.288 KHz, and 0.286 KHz. (These are actually mixed in transmission to narrow the range of modulated frequencies.) The satellite receives and retransmits the pulse containing these measurement frequencies on two additional carrier waves of 449 MHz and 224.5 MHz. The phase of the returned measurement signals is compared with those of the outgoing signals. The observed phase shifts are functions of the distance traveled.

If the distance from the satellite to the ground station is ρ and the wavelength of the measurement frequency is λ, then the number of wavelengths in the round trip is $2\rho/\lambda$ (assuming that ρ is an even number of λ). The distance ρ to a moving satellite would normally be comprised of several whole wavelengths and fractional wavelengths. This fractional wavelength corresponds to the measured phase shift. If expressed in radians, the phase shift is

$$\phi = \frac{4\pi\rho_\phi}{\lambda} \quad \text{or} \quad \rho_\phi = \frac{\phi\lambda}{4\pi}$$

where ρ_ϕ is the range corresponding to the phase shift. (Caution: do not confuse ϕ and λ with geodetic latitude and longitude.)

It is apparent that if only ϕ is measured and the distance corresponding to a fraction of λ computed, the total distance has not necessarily been determined; in other words, the total number of cycles has not been counted. This problem and its solution will be illustrated in the following paragraphs.

Step 1: compute the half wavelengths of the measurement frequencies (the half wavelength is employed since only half of the total distance traveled is sought). Recalling the relationship between wavelength and frequency, $\lambda = C/f$, where C is the propagation velocity in vacuo, the half wavelengths and frequencies for measurement are:

f		$\lambda/2$
f_1 (very fine)	585.533 KHz	256 m
f_2 (fine)	36.596 KHz	4096 m
f_3 (coarse)	2.288 KHz	65,536 m
f_4 (very coarse)	0.286 KHz	524,288 m

Step 2: Assume that ϕ_{f_4}, the phase shift for very coarse frequency 4, results in a distance of 456, 704 m. Without additional knowledge, it is not known whether this is correct or if $\lambda_4/2$ should be added. With SECOR, this problem is resolved by an approximate extended range measurement which is the product of transit time rather than phase shift. In this example, the extended range indicated a distance of about 1,000,000 m; therefore, the very coarse range is

$$\left(1 + \frac{\phi_{f_4}}{2\pi}\right)\frac{\lambda_4}{2},$$

that is,

$$
\begin{array}{r}
456,704 \\
+524,288 \\
\hline
980,992 \text{ m.}
\end{array}
$$

Because range resolution with phase shifts of long waves does not yield the required precision, the foregoing process is repeated with ϕ_{f_3}. The measured phase shift produces 1792 m. It is determined that $\lambda_3/2$ goes into 980,992 about 15 times; therefore, the coarse range is

$$\left(15 + \frac{\phi_{f_3}}{2\pi}\right)\frac{\lambda_3}{2},$$

that is,

$$
\begin{array}{r}
15 \times 65,536 = 983,040 \\
+ \quad 1,792 \\
\hline
+984,832 \text{ m.}
\end{array}
$$

In turn, the fine range is

$$\left(240 + \frac{\phi_{f_2}}{2\pi}\right)\frac{\lambda_2}{2},$$

that is,

$$240 \times 4{,}096 = 983{,}040$$
$$+ \quad 1{,}080 \text{ (measured)}$$
$$\overline{+984{,}120 \text{ m.}}$$

The very fine range is

$$\left(3{,}844 + \frac{\phi_{f_1}}{2}\right)\frac{\lambda_1}{2},$$

that is,

$$3{,}844 \times 256 = 984{,}064$$
$$+ \quad\quad 018 \text{ (measured)}$$
$$\overline{984{,}082 \text{ m.}}$$

With the $\lambda/2$ of the very fine measuring frequency and ability to resolve phase shifts to six milliradians, this final range is seen to have a resolution of 0.25 m.

This observed range is greater than the true range and therefore must be corrected for delays in the ground system and satellite, for tropospheric refraction, and for effects of the ionosphere [14].

Delays in the satellite appear to remain constant and may be determined prior to launch. The corrections thus applied are estimated to have a probable error of ± 1 m. Delays in the ground station may be determined with a test transponder before and after each satellite pass. Such calibration yields a correction with an error estimate of ± 2m.

The tropospheric refraction occurring in the lower atmosphere is removed through an analytical model and surface measurements. The error estimated for this correction is between ± 0.25 m for high elevation angles and ± 1 m for low elevations.

The ionosphere, which extends from 60 km to 1000 km, also causes the observed range to be greater than the true range. This effect is inversely proportional to the frequency of the carrier wave. This fact is used to determine the correction. Recall that the satellite retransmits the measurement frequencies on two different carrier waves. The difference in the ranges calculated for these two waves is the basis for the correction. The probable error estimated for this correction varies between ± 0.25 m for high elevations during night observations to ± 6 m at low elevations during the day.

An additional error of ± 1 m is caused by random noise and by uncertainties in frequency and propagation velocity.

If these error estimates are root-sum-squared by category (high, low, day, night), the error of a single range measurement is found to vary between ± 2.7 m and ± 6.9 m.

In order to solve the geodetic problem presented in the section on intervisible range observations, three known stations and one unknown station must measure range simultaneously on at least three satellite positions. As there is only one transponder in the satellite, this simultaneous ranging from four stations is effectively achieved by time sharing (sequential collation of range). Interrogation and reception by each station is completed in 12.5 milliseconds. All four stations complete a range measurement every 50 milliseconds. The first station to "see" the satellite is designated the "master" station. This station emits a timing signal which is retransmitted from the satellite to all stations. This signal starts the sequence of interrogation.

Through electronic damping, a smooth and continuous picture of phase shift is maintained at each station. On receipt of the master time signal, each station records this phase shift. In this way, simultaneous ranging is effected. For a 7-minute satellite pass, 8400 observations are recorded per station.

The SECOR ground station complex consists of three shelters as

Figure 10–13. SECOR ground station. From left to right: storage shelter, data handling shelter, and radio frequency shelter (Cubic Corporation).

shown in Fig. 10–13. The radio frequency shelter contains the transmitter and receiver, power supply, and antenna controls; this unit also supports the antenna that tracks the satellite. The data handling shelter houses the magnetic tape recorders for time and range, control console, and timing unit. This unit controls modulation and timing of measurement signals and converts the signals received from the satellites for recording on magnetic tape. The storage shelter contains radio communications and test and maintenance equipment [8].

Tests of this system in the simultaneous mode were conducted in the United States in 1964. A small quadrilateral of approximately 500 miles between stations was formed with SECOR stations at Stillwater, Oklahoma; Las Cruces, New Mexico; Austin, Texas; and Fort Carson, Colorado (the unknown station). Solutions were computed for multiple passes. The mean position resulting from this test agreed with the survey position as follows (note that the survey positions are not without error): latitude, 0.7 m; longitude, 7.1 m; geodetic height, 5.7 m. The precision of the solution, the root-mean-square (see Chapter 11) of the difference between individual solutions and the mean position, was: latitude, ±3.5 m; longitude, ±3.9 m; and geodetic height, ±2.4 m.

A similar test was conducted for a larger quadrilateral with sides of approximately 1200 miles. The solution resulted in almost the same precision but the disagreement with survey coordinates was larger: latitude, 14.4 m; longitude, 10.6 m; and geodetic height, 9.2 m [8].

The preceding test results indicate a relative accuracy of one part in 100,000. It is to be anticipated that improved accuracy will be achieved with experience.

The Tokyo Datum was recently connected to the North American Datum through a series of SECOR quadrilaterals across the Pacific. The reported precision of the Tokyo Datum coordinates determined for the islands of one quadrilateral was [14]:

	Latitude (")	Longitude (")	Geodetic Height (m)
Minami-diato-Shima	1.6	1.6	1.0
Iwo Jima	2.1	2.2	2.8
Marcus	3.7	2.2	0.5
Guam	1.3	1.5	2.7

These precision figures were obtained from internal agreement of the solution. They are better than those of the test quadrilateral and this is attributable to more satellite observations in the solution.

Range-Rate Instruments. The Transit Network (TRANET) system sponsored by the U.S. Navy illustrates these doppler instruments and techniques for geodesy. Although the initial intent was the development of a navigation system, the tests conducted over the past several years have shown this approach to be a valuable tool for geodetic positioning on a global datum and in defining the geopotential. Very sophisticated and comprehensive orbit models have resulted from this effort.

This technique has in common with the camera approach the fact that the ground sensor does not interrogate the satellite. The sensor merely observes satellite activity and, therefore, the system cannot be saturated.

Commensurate with the navigational intent, there are different types of satellites available. For navigation, the satellite must provide orbit information. As the presentation at hand is concerned with geodesy, our discussion will be limited to doppler observations of the GEOS A satellite.

The satellite transmits a continuous signal at 162 MHz, 324 MHz, and 972 MHz. Additionally, time marks are transmitted from the satellite every minute. These time hacks consist of a 0.3-second burst of phase modulation and are in synchronization with WWV to 0.4 millisecond or better. By maintaining the time standard in the satellite, the problem of coordinating time between sensors of this system is virtually eliminated.

The ground station receives this signal and differences it with a reference frequency in the receiver;

$$F_0{}^- = F_K - F,$$

where F_K is the reference frequency and F is the received frequency and is a function of the satellite transmitted frequency and the doppler shift, as explained in the development of the doppler mathematics.

The received frequency F must be corrected for ionospheric and tropospheric refraction. The dual frequencies, 324 MHz and 162 MHz, are employed as in the case of SECOR to obtain the ionospheric correction.

The ground receiver counts the number of cycles, N_R, of F_0 occuring within a given time interval. The integrated count is made on the satellite-transmitted time signal. The count, time, and refraction data are recorded and stored for subsequent reduction.

The offset frequency, $F_K - F_T$, is normally between 16 and 32 KHz, and the doppler frequency is approximately ± 10 KHz.

In essence, this solution depends upon stable oscillators in the satellite and ground stations for time increments and frequency control, an epoch time, corrections for refraction, and the satellite position. The error in geodetic position contributed by uncertainties in these factors for one pass has been estimated as follows [7]:

Meters (rms)

Random error (instrumentation)	8
Timing errors	3
Tropospheric refraction errors	3
Ionospheric refraction errors	3
Orbit errors	75
	—
rss	76

Disregarding orbit uncertainties, the system accuracy is consistent with that estimated for SECOR. Observations of multiple satellite passes entered in the solution will greatly reduce the contribution of random errors.

The largest error source stems from uncertainty in the satellite position. This is a subject of some disagreement. Total positional accuracy resulting from multiple passes as calculated in 1965 [2] was 25 m; however, the estimates of 1968 were one-half this value.

In this general orbital solution, the TRANET consists of 12 fixed tracking stations, portable vans, a satellite control center (SCC), and two computing facilities. The purpose of the fixed stations, SCC, and computing facilities is to determine precise orbit information from doppler observations and to monitor the satellites. This effort, combined with observations from the mobile vans, has served to position isolated locations on a global geodetic datum. Some of these isolated locations reported as positioned are Greenland, Australia, Japan, Canton Island, Iwo Jima, Okinawa, Yap Island, Wake, Guam, Kwajalein, Eniwetok, Midway, Philippines, Samoa, Marcus Island, Johnston Island, Hawaii, Seychelles, Ascension, and South Africa.

An early test conducted in the United States and Alaska compared chord distance between stations computed from ground survey coordinates and doppler-produced positions. This test showed a general agreement of 15 m.

In the intervisible (translocation) mode, tests conducted between Baltimore, Maryland, and Quantico, Virginia (a distance of 45 miles) yielded a mean result from 23 satellite passes that agreed with survey coordinates within five meters [17]. This is a relative accuracy of 1 :15,000.

Figure 10–14 shows a TRANET van. This unit contains a receiver, stable oscillator, recording device, monitoring equipment, and communications. A smaller unit, designated a "Geoceiver," has been proposed; since this unit promises to be transportable by two men on foot, it could be a valuable addition to the geodetic bag of tools.

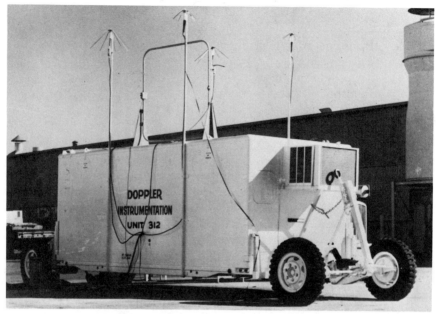

Figure 10–14. Doppler instrumentation van (official U.S. Navy photograph).

SATELLITES

To this point, the method of observing satellites and how such observations are put in use have been presented. The key item in this process, the satellite, will now get some attention.

First, the satellite must have a means of communication with the observer. Reflecting satellites as the Echo series represent the simplest approach to this requirement. PAGEOS is no more than an aluminum mylar balloon. Actually, any object that reflects light is satisfactory and, theoretically, many orbiting objects are available. Of course, in the intervisible mode, all the observing cameras must be accurately time-synchronized, e.g. the shutter action of participating cameras must coincide within at least 0.001 second [16].

On the other extreme, there are satellites like GEOS which communicate with a number of ground-based geodetic systems. The GEOS B spacecraft, shown in Fig. 10–15 atop a Thor-Delta, just prior to liftoff, carries two C-band transponders, a doppler beacon, four flashing lights, laser reflectors, a SECOR transponder, and a range and range-rate transponder. Additionally, it carries a gravity-gradient attitude control subsystem, a power subsystem, a telemetry subsystem, a command subsystem, an optical memory and control unit, and an antenna subsystem. The weight of the spacecraft is 468 pounds. The National

Figure 10–15. GEOS B spacecraft atop Thor-Delta rocket prior to launch (National Aeronautics and Space Administration).

Aeronautics and Space Administration launched the GEOS B with a Thor-Delta rocket, shown in Fig. 10–16, on the Air Force Western Test Range, Vandenberg Air Force Base, California, on January 11, 1968. The orbit achieved has an 850 nautical mile apogee, a 583 nautical mile perigee, an inclination of 105°.8 and a period of 112.184 minutes. This is by far the most sophisticated geodetic satellite ever to be placed into orbit and promises to provide a wealth of data.

The satellite must be in the proper orbit for the job at hand. For example, intervisible observations require that the inclination be at an angle that will provide observations within the latitude limits of interest. The altitude selected is the lowest that will permit simultaneous observations at a given minimum elevation angle. Specifying the orbit

Figure 10–16. Liftoff of GEOS B on the Air Force Western Test Range (National Aeronautics and Space Administration).

for geopotential studies is far more complex. Generally, a variety of orbits should be provided for this undertaking [13].

Table 10–1 is a representative list of geodetic satellites and indicates satellite type, orbit, and launch data.

CONCLUSION

To conclude this discussion of satellite geodesy, several significant items will be repeated for emphasis.

First, all observation techniques presented are independent of the local vertical. Cameras observe directions in inertial space, while the azimuth and elevation angles observed classically with theodolites are

TABLE 10–1

Major Geodetic Satellites

			Initial Orbit			
Satellites	Launch date	Period (min)	Apogee (n.m.)	Perigee (n.m.)	Inclination (deg.)	Geodetic system
Echo 1	12 Aug 60	118.2	914	817	47.2	Reflective
ANNA 1B	31 Oct 62	107.8	632	582	50.1	Flashing lights Doppler beacon SECOR transponder
Echo 2	25 Jan 64	108.8	709	558	81.5	Reflective
Explorer 22	9 Oct 64	104.7	581	477	79.7	Laser reflectors Doppler beacon
Explorer 27	29 Apr 65	107.8	711	507	41.2	Laser reflectors Doppler beacon
SECOR 5	10 Aug 65	122.1	1305	610	69.2	SECOR transponder
GEOS A	6 Nov 65	120.3	1228	602	59.4	Flashing lights Laser reflectors Doppler beacon SECOR transponder Range and range-rate transponder
PAGEOS	23 June 66	181.4	2312	2264	87.1	Reflective
SECOR 9	22 Jun 67	172.1	2128	2051	89.8	SECOR transponder
GEOS B	11 Jan 68	112.2	850	583	105.8	Flashing lights Laser reflectors Doppler beacon SECOR transponder Range and range-rate transponder Radar transponder

referred to the local direction of gravity. Of course, direct measurements of distance and distance differences do not depend on the direction of the vertical unless such quantities are to be reduced to a reference ellipsoid.

In the simultaneous mode, only the coordinate difference between the known and unknown sites is determined; that is, only relative position is determined. In the orbital mode, the position produced refers to a CG-centered coordinate system.

The shape of the reference ellipsoid is determined only by analysis of the satellite orbit and the resulting coefficients that define the geopotential and gravity.

In successful operations reported to date, TRANET has been employed primarily in the orbit mode while the cameras and SECOR have made their marks with simultaneous observations. Reversal of these roles as well as increased use of short-arc techniques may be seen in the future.

Table 10–2 summarizes the important aspects of the systems discussed.

TABLE 10–2

Geodetic Satellite Observing Systems

	Cameras	SECOR	TRANET
Observations:	Directions	Range	Range rate
Observation Accuracy (one sigma):	$\pm 1''$ for PC–1000	± 3 to ± 5 m	± 5 cm/sec
	$\pm 2''$ for BC–4		
Solution with Orbit Known:			
Required Observing Stations	1	1	1
Minimum Satellite Observations	2	3	3
Position Determined	Refers to the center of gravity		
Solution with Orbit Unknown:			
Required Known Observing Stations	2	3	1
Maximum Unknown Stations Allowed	Unlimited	1	Unlimited
Position Determined	Refers to coordinate system of known stations		

REFERENCES

1. Aeronautical Chart and Information Center, Determination of Ground Positions from Observations of Artificial Earth Satellites, ACIC TR 86, October 1959.
2. R. J. Anderle, Geodetic parameters set NWL 5 E–6 based on doppler satellite observations, U.S. Naval Weapons Laboratory Report No. 1978, April 1965.
3. R. M. L. Baker, Jr. and Maud W. Makemson, *An Introduction to Astrodynamics*, Academic Press, New York, 1960.
4. J. H. Berbert, Intercomparison of GEOS A observation systems, NASA–TM–X–55945, National Aeronautics and Space Administration, Goddard Space Flight Center, Greenbelt, Maryland, July 1967.
5. D. Brown, A treatment of analytical photogrammetry with emphasis on ballistic camera applications, RCA Data Reduction Technical Report No. 39, 1957.
6. C. E. Ewing, Research and development in the field of geodetic science, Air Force Surveys in Geophysics 124, AFCRL–TN–60–435, Bedford, Massachusetts, August 1960.
7. W. H. Guier, R. R. Newton, and G. C. Weiffenbach, Analysis of the observational contributions to the errors of the Navy satellite doppler geodetic system, APL/JHU Report TG–653, January 1965.

8. T. J. Hayes, SECOR for Satellite Geodesy, presentation to the Tenth International Congress of Photogrammetry, Lisbon, Portugal, September 1964.

9. W. A. Heiskanen, Intercontinental connection of geodetic systems, Ohio State Research Foundation Technical Paper, Columbus, Ohio, 1955.

10. D. N. Huber and N. Donovan, The Air Force PC–1000 Chopping Shutter Camera System, presentation to the 1967 Semi-Annual American Society of Photogrammetry/American Congress of Surveying and Mapping Convention, St. Louis, Missouri, October 1967.

11. W. M. Kaula, Celestial Geodesy, National Aeronautics and Space Administration, Technical Note D–1155, Goddard Space Flight Center, Greenbelt, Maryland, March 1962.

12. R. B. Kershner, Present state of navigation by doppler measurement from near earth satellites, *APL Technical Digest*, November–December 1965.

13. I. I. Mueller, *Introduction to Satellite Geodesy*. Frederick Ungar Publishing Co., Inc., New York, 1964.

14. N. J. D. Prescott, Experience with SECOR Planning and Data Reduction, presentation to International Association of Geodesy Symposium of Electronic Distance Measurements, Oxford, England, September 6–11, 1965.

15. M. S. Tavenner, LARGOS: A Suggested Method for Stereo Triangulation, A Compendium of Papers in the Fields of Geodesy and Planetary Geometry prepared at AFCRL during 1962, AFCRL 63–876, August 1963, O. W. Williams, Edit.

16. E. A. Taylor, Optical tracking system for space geodesy, *Proc. First International Symposium on the Use of Artificial Satellites for Geodesy*. Washington, D.C., April 26–28, 1962; North-Holland Publishing Co., 1963, G. Veis, Edit.

17. E. E. Westerfield and G. Worsley, Translocation by navigation satellite, *APL Technical Digest*, July–August 1966.

18. I. D. Zhongolovich, Earth satellites and geodesy, *Soviet Astronomy*, **41**, No. 1 (July and August 1964); NASA TT F–8868.

Chapter 11 Adjustment Computations

The geodesist must be not only concerned with position, azimuth, and distance, but also with the adequacy of such determinations. In the pursuit of the general geodetic objective of determining the size and shape of the earth, accuracy estimates associated with measurements obviously indicate how well the geodesist is doing. Data adequacy must be known even to approach some tasks. In the case of satellite observations, not only must a radar provide position, time, and velocity for the satellite track, but also a statement as to how well these represent the truth is mandatory in order to combine these data with others in defining the orbit. To provide such a statement, the numerical uncertainties in the radar observation and its geodetic position must be known if the measurements are to be useful—one without the other is meaningless.

In either case, accurate measurements are required. How do we get an accurate measurement? One way, obviously, is to measure the quantity several times. Unfortunately, this results in several answers for a single quantity. Now, which answer is the best and how close is it to the truth?

Error is the difference between an accepted value and the truth. Rarely, if ever, is the true error in a measurement known. (If the true error were known, the true value would also be known.)

The problems to be probed are:

1. Errors that cause survey determinations to vary from the truth. Methods are found to represent precision (repeatability) and approximate accuracy (truth).

2. Selection of the best answer when many are available, and a judgment of the adequacy of this selection.

3. The meaning of the accuracy estimated for terms derived.

ERRORS

There are three basic types of error: blunders (more commonly termed "operator or human error"), systematic errors, and random errors. A distance measurement between two posts a kilometer apart will be used to illustrate these types of error. This distance is measured in 20 increments with a 50-meter metal tape.

Blunders are simply mistakes or goofs. Because the source of such errors are generally human, they cannot be predicted. In the sample task, suppose the surveyors lost count of the number of tape lengths between the posts and recorded 19 instead of 20. The recorded distance, the sum of the tape lengths, would be 950 meters. This distance varies from the truth because of a 50-meter blunder.

The effect of blunders is usually eliminated by repeated measurements. In this case, it is likely that during remeasurement of the distance the same error would not be made, and certainly even less likely during a third run. Such errors are normally large and are readily identified.

Systematic errors are errors resulting from a predictable source. For instance, suppose air temperature were high and a survey tape expanded (lengthened). Although the tape read 50 meters, it might be 50.01 meters in length. In this case, approximately 19.9960 tape lengths would exist between the posts, and the total distance computed on the assumption of a 50-meter tape would be 999.800 meters. Repeated measurements under the same condition would yield the same results and would not indicate error (a case of precision but not accuracy). However, if the tape expansion were known as a function of temperature, a correction factor could be applied.

Systematic errors are typically of constant magnitude and size under any one set of given conditions. The source of such errors is generally the measuring instrument. When the cause is understood, such effects can be removed. Procedure or computations instead of repetition corrects systematic errors.

Random errors are those that remain after blunders have been removed and systematic errors have been corrected. Such errors can only be defined by repeated measurements. They indicate how well a measurement may be repeated and characterize the precision of a measurement. If they are the only error source, they are a measure of accuracy. By definition, random errors are small and are as often positive as negative.

For example, suppose the 1-kilometer line were measured six times with the following results:

$$
\begin{array}{r}
1000.018 \text{ meters} \\
1000.052 \\
999.983 \\
999.967 \\
999.954 \\
1000.049 \\
\hline
6000.023 \text{ meters.}
\end{array}
$$

Which of these measurements is the correct one? What figure should be used to represent the distance between the posts? In a situation like this, common sense says that the mean value is the most probable, that is,

$$\text{Mean} = \frac{6000.023}{6} = 1000.004 \text{ meters.}$$

The questions now arise, "What chance does this figure have of representing the truth? What is the accuracy?" (We know that it is in error by +4 mm; however, this is the type of inside information one is not normally privileged to possess. By literary definition, the line is 1 kilometer in length.)

The following symbols are necessary for the exploration of the mean value and its associated accuracy estimates:

$$T = \text{true value,}$$
$$l_i = \text{observed values,}$$

where $i = 1, 2, 3, 4, \ldots n$,

$$n = \text{number of observations,}$$
$$E_i = \text{true error of } l_i; \quad E_i = T - l_i.$$

The *mean* value approximating T is the sum of l_i divided by the number of measurements,

$$L = \frac{[l]}{n}. \tag{11–1}$$

(Throughout this chapter, the square brackets, [ ], in equations indicate summation.) The difference between the mean and observed value is termed the *residual* and is expressed as

$$v_i = L - l_i. \tag{11–2}$$

Obviously,

$$[v] = nL - [l] = n\,\frac{[l]}{n} - [l],$$

$$[v] = 0. \tag{11–3}$$

Furthermore, to substantiate the use of the mean as the best choice, note that

$$[v^2] = v_1{}^2 + v_2{}^2 + v_3{}^2 + \ldots v_n{}^2$$
$$= (L - l_1)^2 + (L - l_2)^2 + \ldots (L - l_n)^2$$
$$= nL^2 - 2L\,[l] + [l^2];$$

and, with Eq. (11–1),

$$[v^2] = \frac{[l]^2}{n} - 2\,\frac{[l]^2}{n} + [l^2],$$

$$[v^2] = [l^2] - \frac{[l]^2}{n}. \tag{11–4}$$

·If, on the other hand, some value other than the mean were selected —R, for instance—the result is:

$$r_1 = R - l_1, r_2 = R - l_2, \text{ etc.}$$

Now,

$$[r^2] = (R - l_1)^2 + (R - l_2)^2 \ldots$$
$$= nR^2 - 2R [l] + [l^2].$$

Substituting the expression for $[l^2]$ extracted from Eq. (11–4) yields

$$[r^2] = nR^2 - 2R [l] + [v^2] + \frac{[l]^2}{n}$$

$$= [v^2] + n\left(R - \frac{[l]}{n}\right)^2.$$

As pointed out in [3], it is evident that $[r^2]$ is greater than $[v^2]$; hence, the mean value, L, yields a minimum sum of residuals squared. The concept of

$$[v^2] = \text{minimum}$$

is basic to all adjustments discussed. The goal is to achieve the most probable answer with minimum alterations to the observations.

The true error of the mean is expressed as

$$E_L = T - L$$

and, by definition, the standard error of one observation, l_i, is

$$m = \left(\frac{[E^2]}{n}\right)^{1/2}. \tag{11–5}$$

The true error of a single observation, E_i, may be written by substituting the expression for E_L, and from Eq. (11–2):

$$E_i = T - l_i = E_L + L - l_i = E_L + v_i.$$

With reference to Eq. (11–5),

$$nm^2 = [E^2] = nE_L^2 + 2E_L [v] + [v^2];$$

and as $[v] = 0$,

$$nm^2 = nE_L^2 + [v^2].$$

Since the true error of the mean, E_L, is never known, it is approximated by the *standard error of the mean*, which is defined as

$$M = \frac{m}{(n)^{1/2}} \doteq E_L. \tag{11–6}$$

(This expression is derived later.) On substitution of this approximation,

$$nm^2 = m^2 + [v]^2,$$

$$m = \left(\frac{[v^2]}{n-1}\right)^{1/2}. \tag{11–7}$$

This is the *standard error of a single measurement* which formed part of a mean determination. In the example of the six measurements of the 1-kilometer line, the residuals would be computed with Eq. (11–2), squared, and summed:

$$[v^2] = 0.008835.$$

With Eq. (11–7)

$$m = \left(\frac{0.008835}{6-1}\right)^{1/2} = \pm 0.04204 \text{ meter.}$$

Furthermore, from Eq. (11–6), the standard error of the mean value is found to be

$$M = \frac{0.04204}{(6)^{1/2}} = \pm 0.017 \text{ meter.}$$

With all systematic errors removed, M indicates how well L can be expected to represent T (within a certain probability which will be defined later).

The m indicates how well this measurement can be repeated—the measurement precision. If and only if all systematic errors, biases, and blunders have been removed can m be equated with accuracy.

The standard error of an observation as well as the equivalent error for its mean value have been presented. This issue is completed by extending the process to the general case of error estimates of functions, that is, when $C = f(A, B)$, what is m_C?

Consider the simple function

$$C = A + B,$$

where A and B are measured n times. If E_{Ai} designates the true error in A and E_{Bi} the true error in B for the ith measurement, then

$$E_{Ci} = E_{Ai} + E_{Bi},$$

and, on squaring,

$$E_{Ci}^2 = E_{Ai}^2 + E_{Bi}^2 + 2E_{Ai}E_{Bi};$$

summing these expressions for n determinations of A and B yields

$$[E_C{}^2] = [E_A{}^2] + [E_B{}^2] + 2[E_AE_B].$$

Dividing this expression by n and using Eq. (11–5), we obtain

$$m_C{}^2 = m_A{}^2 + m_B{}^2 + 2\frac{[E_AE_B]}{n}.$$

The term $[E_AE_B]/n$ is designated *covariance* and is equal to zero providing A and B are independent variables; that is, the determination of one is in no way influenced by the determination of the other. Or, since positive and negative random errors are of equal probability, it can be assumed that the same trend applies to E_AE_B. Providing this trend persists, these terms would cancel in summation. The covariance will be treated as zero in the following calculations. With this assumption, the equation for the *standard error of sums or differences* becomes (called the root-sum-square or simply rss)

$$m_C = (m_A{}^2 + m_B{}^2)^{1/2}, \tag{11-8}$$

and, in the special case when $m_A = m_B$, then

$$m_C = m(n)^{1/2}, \tag{11-9}$$

where n refers to the number of terms summed. Because m can be $+$ or $-$, it is apparent that a simple algebraic summation would be difficult to apply:

$$m_C = \pm m_A \pm m_B = \ ?$$

In the example of the 1-kilometer line, suppose that past experience gives the standard error (random) of one tape length as 0.009 meter. With this value and Eq. (11–9), the standard error to be expected for one measurement of the line would be computed as

$$m = 0.009\ (20)^{1/2} = \pm\ 0.0402 \text{ meter.}$$

Similarly the standard error for one tape length can be computed from the standard error of one line measurement contributing to the mean value by using residuals, v, and Eq. (11–7):

$$m_C = \left(\frac{[v^2]}{n-1}\right)^{1/2} = \pm\ 0.04204 \text{ meter.}$$

Therefore, from (11–9),

$$m = \frac{m_C}{(n)^{1/2}} = \frac{0.04204}{(20)^{1/2}} = \pm 0.009 \text{ meter.}$$

Now consider the standard error of a quantity that is the function of the product of observations, that is: $C = A \times B$. If $C = Kl$, where K is a constant and l is observed, the true error of C would be $E_C = KE_l$. In the case under consideration, where both terms are observed, the error equation is $E_C = BE_A + AE_B$. Again, squaring and summing the n determinations of A and B yields

$$[E_C{}^2] = B^2[E_A{}^2] + A^2[E_B{}^2] + 2AB\ [E_A E_B].$$

The standard error for C is found by dividing this expression by n, assuming the last term to be zero (assuming A and B to be independent of one another), and by using Eq. (11–5):

$$m_C = (B^2 m_A{}^2 + A^2 m_B{}^2)^{1/2}. \tag{11-10}$$

Now, based upon the foregoing derivations of Eqs. (11–8)–(11–10), a general expression for the *standard error of a function* using partial derivatives may be written. For $C = f(A, B)$,

$$m_C = \left[\left(\frac{\partial f}{\partial A} m_A\right)^2 + \left(\frac{\partial f}{\partial B} m_B\right)^2\right]^{1/2}. \tag{11-11}$$

To use Eq. (11–11), consider the expression for the mean, L, resulting from multiple observations, l_i, with equal standard error, m.

$$L = \frac{l_1 + l_2 + l_3 + l_4 + \ldots l_n}{n} = f(l).$$

As $\partial f / \partial l = 1/n$ and $m_1 = m_2 = m_3 = m_4$, etc.;

$$m_L{}^2 = n \left(\frac{m}{n}\right)^2 = \frac{m^2}{n},$$

or

$$M = \frac{m}{(n)^{1/2}}. \qquad (11\text{–}6)$$

These equations deal with purely random errors. It should be understood, however, that these random errors may be combined with errors of a nonrandom complexion. Referring again to the example, suppose the tape had been calibrated and certified to be 50 ± 0.001 meters. This is not the repeatability stated previously, i.e., the ± 0.009 meter; this means that the true tape length of 50 meters is always ± 0.001 meter in error. Although the sign of this error is unknown, the 0.001 error is constant and is not handled as a random error. If the random error is designated m and this calibration uncertainty c, these would be combined as

$$m_C = \pm (m(n)^{1/2} + cn).$$

Understanding the difference between these uncertainties is mandatory in making such error estimates.

Weighting the Error. The important subject of *weights* can no longer be ignored. Because all measurements forming a solution may not be of the same adequacy, this subject must be considered. Suppose the 1-kilometer line were measured on a second day, and the mean length, L_2, were computed from 3 measurements rather than from 6. The standard error of this mean would be

$$M_2 = \pm \frac{m}{(3)^{1/2}} = \pm \frac{0.04204}{(3)^{1/2}} = \pm 0.0243 \text{ meter.}$$

(The same standard error for one measurement, m, is applied.) Since the original mean, L_1, had a standard error of ± 0.017 meter, L_1 and L_2 obviously are not of the same quality. It is not appropriate to determine an improved length with a second simple mean. This problem is resolved by assigning weights to these values commensurate with their relative worth. With p designating weight,

$$L_p = \frac{p_1 L_1 + p_2 L_2}{p_1 + p_2} = \frac{[pL]}{[p]}. \qquad (11\text{–}12)$$

In the example, $p_1 = n_1 = 6$ and $p_2 = n_2 = 3$; therefore, the weighted mean is

$$L_p = \frac{6L_1 + 3L_2}{9}.$$

This relationship is easily verified by substitution of the original expressions for L_1 and L_2:

$$L_p = \frac{\dfrac{6(l_1 + l_2 + l_3 \ldots l_6)}{6} + \dfrac{3(l_7 + l_8 + l_9)}{3}}{9},$$

$$= \frac{(l_1 + l_2 + l_3 + l_4 \ldots l_9)}{9} = \frac{[l]}{n}.$$

By substituting $p_1 + p_2$ for n in Eq. (11–6), the *standard error of the weighted mean* is simply expressed as

$$M_p = \pm \frac{m}{([p])^{1/2}}. \tag{11–13}$$

Backtracking further, the standard errors of L_1 and L_2 are written as

$$M_1 = \frac{m}{(p_1)^{1/2}} \quad \text{and} \quad M_2 = \frac{m}{(p_2)^{1/2}}.$$

Squaring and equating these expressions yield

$$\frac{M_2{}^2}{M_1{}^2} = \frac{p_1}{p_2}. \tag{11–14}$$

It is seen that weights may be expressed as ratios that are inversely proportional to the squares of the errors. In the application of Eq. (11–14), the weight corresponding to the least accurate measurement could be assigned a value of 1 (a unit weight) and then p_1 computed accordingly. In the example, let $p_2 = 1$; therefore,

$$\frac{p_1}{1} = \frac{M_2{}^2}{M_1{}^2} = \left(\frac{.024}{.017}\right)^2 = 2,$$

which is consistent with the weight ratio used,

$$\frac{p_1}{p_2} = \frac{6}{3} = 2.$$

To this point, all observations have been considered of equal validity and error expressions were derived accordingly. In reality, however, Eq. (11–7) must be written

$$m = \left(\frac{[pv^2]}{n-1}\right)^{1/2}$$

where p is the weight of respective observations that formed the v. The m is the standard error for an observation of unit weight.

LEAST SQUARES

The subject of adjustments was introduced by using multiple observations of a single unknown. It was shown that the mean of such observations yields a minimum value for the sum of the residuals squared—the residuals being the difference between the observed values and the mean one. In this sense, the residual may be thought of as the error of the observation compared with the mean.

Clearly, from the foregoing development, the mean of several observations yields a reasonable solution for a single unknown. However, what is done if there is more than one unknown and these are not observed directly? A simple arithmetic mean does not seem applicable to this situation.

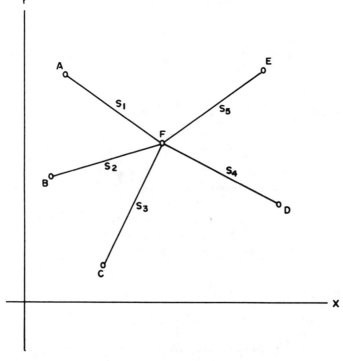

Figure 11–1. Distance observations for solution.

Consider the case in Figure 11–1 which depicts observed distances from five known points to an unknown point, F, on a two-dimensional coordinate system. Here there are five observations—S_1, S_2, S_3, S_4, and S_5—and two unknowns—X_F and Y_F. Obviously, plane trigonometry permits any two of these distances to be used to solve for the two unknowns. In fact, with all five sides employed, four different, independent solutions are possible for the two coordinates. These solutions will not agree because of errors in the measured distances. Again, we have the problem of known quantities outnumbering the unknowns, that is, $n > U$. The question is asked again, "Which solution is the correct one?"

The method of least squares offers an answer to this question. The theory behind this method is over 170 years old and is attributed to the genius of Carl Friedrich Gauss. Gauss developed this concept while a 17-year old student at the University of Göttingen. This method can be used to determine the most probable value of unknowns that are functions of excessive observations and to estimate the error of the solution as well as that of the observations. The problem was created because $n > U$; the least squares technique applies only under this condition.

Within this theory, the methods of *variation of parameters* and then that of *conditions* will be explored.

Variations of Parameters. The initial development of the variations of parameters concept will refer to Figure 11–1 in order to place meat on some abstract bones. The symbols used are:

l_i = observed values,
l_i^0 = computed values with approximate unknowns,
l_i = observed minus computed = $l_i - l_i^0$,
L_i = adjusted value of observations,
v_i = correction to observation = $L_i - l_i$,
X^0, Y^0 = approximate unknowns,
x, y = correction to approximate unknowns,
X, Y = adjusted unknowns = $(X^0 + x), (Y^0 + y)$.

(The subscript i represents the observation lines: $i = 1, 2, 3, 4, \ldots n$; or known stations: $i = A, B, C, D, \ldots n$.)

First, estimate the observed values with the approximate unknowns,

$$l_i^0 = f(X^0, Y^0). \tag{11-15}$$

For the circumstances illustrated in Figure 11–1, the line lengths are computed by writing Eq. (11–15):

$$(S_i^0)^2 = (X_i - X^0)^2 + (Y_i - Y^0)^2.$$

Next, obtain a differential equation in terms of alterations to computed observations on the left, dl^0, and to approximate unknowns on the right, x and y. This would have the form

$$dl_i^0 = \frac{\partial f}{\partial X} x + \frac{\partial f}{\partial Y} y;$$

and if
$$L_i - l_i{}^0 = dl_i{}^0,$$
then
$$L_i = l_i{}^0 + \frac{\partial f}{\partial X} x + \frac{\partial f}{\partial Y} y.$$

In this equation, corrections to the approximate coordinates used to compute $l_i{}^0$ are the unknowns. Recalling that the adjusted observation value, L_i, equals the measured value plus a correction, v_i, then

$$L_i = l_i + v_i; \tag{11-16}$$

and symbolizing the difference between the measured and computed values with

$$l_i = l_i - l_i{}^0,$$

the *general observation equation* is written as

$$v_i = a_i x + b_i y - l_i \tag{11-17}$$

where $a_i = \partial f / \partial X$ and $b_i = \partial f / \partial Y$. (If the observations this equation represents are of different validity, weights are assigned. The approach is identical to that discussed in conjunction with weighted means. The square root of the weight, $(p)^{1/2}$, is multiplied through the observation equation and carried in subsequent computations.)

In the example problem,

$$l_i = S_i - S_i{}^0,$$

$$a_i = \frac{\partial S_i}{\partial X} = \frac{X^0 - X_i}{S_i},$$

$$b_i = \frac{\partial S_i}{\partial Y} = \frac{Y^0 - Y_i}{S_i};$$

therefore,

$$v_i = \left(\frac{X^0 - X_i}{S_i}\right) x + \left(\frac{Y^0 - Y_i}{S_i}\right) y - (S_i - S_i{}^0).$$

An equation is written for each observation. In this case there would be five equations and two unknowns.

Each equation is squared, that is,

$$v_i{}^2 = (a_i x)^2 + 2a_i b_i xy - 2a_i l_i x + (b_i y)^2 - 2b_i l_i y + l_i{}^2,$$

and summed to produce

$$[v^2] = [a^2]x^2 + 2[ab]xy - 2[al]x + [b^2]y^2 - 2[bl]y + [l^2]. \tag{11-18}$$

At this juncture, the goal of solution is identical to that of the arithmetic mean; that is, x and y are sought that produce

$$[v^2] = \text{minimum}.$$

From calculus, this condition is known to exist when the first derivative of the function equals zero, that is $d\,[v^2] = 0$, or equivalently, when the partial derivatives are zero, that is,

$$\frac{\partial[v^2]}{\partial X} = 0, \qquad \frac{\partial[v^2]}{\partial Y} = 0.$$

Therefore, with Eq. (11–18),

$$\frac{1}{2}\frac{\partial[v^2]}{\partial X} = [a^2]x + [ab]y - [al] = 0, \qquad (11\text{–}19a)$$

and

$$\frac{1}{2}\frac{\partial[v^2]}{\partial Y} = [ab]x + [b^2]y - [bl] = 0. \qquad (11\text{–}19b)$$

These are *normal equations*, one equation per unknown. Without much difficulty, these equations may be solved for the two unknowns:

$$x = \frac{[b^2]\,[al] - [ab]\,[bl]}{[a^2]\,[b^2] - [ab]^2}, \qquad (11\text{–}20a)$$

and

$$y = \frac{[a^2]\,[bl] - [ab]\,[al]}{[a^2]\,[b^2] - [ab]^2}. \qquad (11\text{–}20b)$$

With these values, the approximate coordinates originally assumed are adjusted:

$$X = X^0 + x, \qquad Y = Y^0 + y. \qquad (11\text{–}21)$$

A single set of coordinates has been obtained. With these coordinates, adjusted observation values, L_i, may be computed and, in turn, the residuals computed with Eq. (11–16). The $[v^2]$ required for an error estimate could be computed from this. A more direct route to $[v^2]$, however, is found by rewriting Eq. (11–18) as

$$[v^2] = ([a^2]x + [ab]y - [al])x + ([ab]x + [b^2]y - [bl])y - [al]x$$
$$- [bl]y + [l^2].$$

Because the first two expressions in parentheses are Eqs. (11–19) and are equal to zero, then

$$[v^2] = -[al]x - [bl]y + [l^2]. \qquad (11\text{–}22)$$

The minimum sum of the residuals squared is thus computed. Now, what form will the standard error take? Recall that in the case of the mean and one unknown, the form was

$$m = \left(\frac{[v^2]}{n-1}\right)^{1/2}.$$

The general form of this expression is

$$m = \left(\frac{[v^2]}{n-U}\right)^{1/2} \qquad (11\text{–}23)$$

and in the example, $U=2$. This is the standard error of unit weight for this adjustment; it is the standard error of a hypothetical observation with $p=1$. Error estimates for all other values deduced from this adjustment are defined in terms of this m.

To determine the *standard errors of the adjusted values*, X and Y, refer back to Eq. (11–11) which expresses the standard error of a function $C=f(A,B)$; that is,

$$m_C = \left[\left(\frac{\partial f}{\partial A} m_A \right)^2 + \left(\frac{\partial f}{\partial B} m_B \right)^2 \right]^{1/2}.$$

Going one step further, express the corrections, x and y, as functions of the l terms and the constants α and β (as yet unspecified):

$$x = \alpha_1 l_1 + \alpha_2 l_2 + \alpha_3 l_3 + \ldots \alpha_n l_n, \tag{11–24a}$$

and

$$y = \beta_1 l_1 + \beta_2 l_2 + \beta_3 l_3 + \ldots \beta_n l_n. \tag{11–24b}$$

Now, from (11–11), it is apparent that

$$m_x{}^2 = \alpha_1{}^2 m_1{}^2 + \alpha_2{}^2 m_2{}^2 + \ldots \alpha_n{}^2 m_n{}^2,$$
$$= [\alpha^2] m^2.$$

Likewise,

$$m_y{}^2 = [\beta^2] m^2.$$

(The standard error, m, is taken as the mean value of m_i.) By referring to Eq. (11–14),

$$m_2{}^2 p_2 = m_1{}^2 p_1,$$

and designating $p_1 = 1$, and substituting m for m_1, it becomes apparent that p_2 or, specifically, p_x and p_y are

$$p_x = \frac{1}{[\alpha^2]} \quad \text{and} \quad p_y = \frac{1}{[\beta^2]}.$$

It is now necessary to express $[\alpha^2]$ and $[\beta^2]$ in the same terms as the observation equations. The solution for x and y is found in Eqs. (11–20). First, look at x:

$$x = \frac{[b^2] [al] - [ab] [bl]}{[a^2] [b^2] - [ab]^2},$$

or, by juggling a bit,

$$x = \frac{[b^2] (a_1 l_1 + a_2 l_2 + \ldots a_n l_n) - [ab] (b_1 l_1 + b_2 l_2 + \ldots b_n l_n)}{[a^2] [b^2] - [ab]^2},$$

$$x = \frac{([b^2] a_1 - [ab] b_1) l_1 + ([b^2] a_2 - [ab] b_2) l_2 + \ldots}{[a^2] [b^2] - [ab]^2}.$$

Comparing this last equation with the form of Eqs. (11–24), we can conclude that

$$\alpha_i = \frac{[b^2]\,a_i - [ab]\,b_i}{[a^2]\,[b^2] - [ab]^2},$$

which, on squaring and adding produces

$$[\alpha^2] = \frac{1}{([a^2]\,[b^2] - [ab]^2)^2}\,([b^2]^2\,[a^2] - 2\,[b^2]\,[ab]^2 + [ab]^2\,[b^2])$$

$$= \frac{[b^2]\,([a^2]\,[b^2] - [ab]^2)}{([a^2]\,[b^2] - [ab]^2)^2}$$

$$= \frac{[b^2]}{[a^2]\,[b^2] - [ab]^2}.$$

Therefore, the relative weight of x resulting from the solution is

$$p_x = \frac{1}{[\alpha^2]} = \frac{[a^2]\,[b^2] - [ab]^2}{[b^2]}.$$

Considering this weight, the *standard error of the adjusted value, x or X,* is

$$m_x = \frac{m}{(p_x)^{1/2}}. \tag{11–25}$$

The same procedure is followed for y and yields

$$p_y = \frac{[a^2]\,[b^2] - [ab]^2}{[a^2]}$$

and

$$m_y = \frac{m}{(p_y)^{1/2}}. \tag{11–26}$$

To complete this discussion, the error estimate for a function whose variables are *dependent* (or correlated) must be considered. This can be illustrated by computing the length of a line with the X and Y coordinates just determined. Since the X and Y are functions of identical observations, they were not solved for independently—if the X were altered, so must be the Y. In the derivations of Eq. (11–11), the $E_x E_y$ was assumed to be zero. In the situation being considered, this is no longer a valid assumption. For this function, $F = f(X, Y)$, Eq. (11–11) is written as

$$m_F = \left[\left(\frac{\partial f}{\partial X} m_x \right)^2 + \left(\frac{\partial f}{\partial Y} m_y \right)^2 + 2\,\frac{\partial f}{\partial X}\,\frac{\partial f}{\partial Y}\,E_x E_y \right]^{1/2},$$

where the E_x and E_y are the true errors of x and y. As with Eq. (11–24),

$$E_x = \alpha_1 E_1 + \alpha_2 E_2 + \alpha_3 E_3 + \dots,$$
$$E_y = \beta_1 E_1 + \beta_2 E_2 + \beta_3 E_3 + \dots,$$

and

$$E_x E_y = \alpha_1 \beta_1 E_1{}^2 + \alpha_2 \beta_2 E_2{}^2 + \dots 2\alpha_1 \beta_2 E_1 E_2 + 2\alpha_2 \beta_1 E_2 E_1 + \dots$$

where the E_1, E_2, etc., are the true errors of observations l_1, l_2, etc. Because l_1 and l_2 are independent, $E_1 E_2 = 0$, and with m substituted as the average value of E_1, E_2, etc., we have

$$E_x E_y = [\alpha\beta]m^2.$$

Now, combine

$$\alpha_i = \frac{[b^2]\,a_i - [ab]\,b_i}{[a^2]\,[b^2] - [ab]^2},$$

and

$$\beta_i = \frac{[a^2]\,b_i - [ab]\,a_i}{[a^2]\,[b^2] - [ab]^2},$$

which, upon multiplication and summation, produce

$$[\alpha\beta] = \frac{1}{p_{xy}} = -\frac{[ab]}{[a^2]\,[b^2] - [ab]^2}$$

and

$$m_{xy} = \frac{m}{(p_{xy})^{1/2}}. \tag{11-27}$$

With this and the expression for m_x and m_y, the standard error of the function is expressed as

$$m_F = m\left[\left(\frac{\partial f}{\partial X}\right)^2\frac{1}{p_x} + \left(\frac{\partial f}{\partial Y}\right)^2\frac{1}{p_y} + 2\,\frac{\partial f}{\partial X}\,\frac{\partial f}{\partial Y}\,\frac{1}{p_{xy}}\right]^{1/2}. \tag{11-28}$$

The cycle of computations for variation of parameters has been completed without mention of the most prevalent type of geodetic observation—directions, or angles. To include this important information, equations similar to Eq. (11–17) must be derived (an apparent regression in our progress). As in the case of distances, the computations are made in a plane coordinate system. (Observation equations could be formed by differentiating equations expressing measured quantities in terms of latitude and longitude; however, this form lacks the simplicity of the plane coordinate system and does not produce a better explanation.)

In order to work in a plane coordinate system, observed directions must be reduced to their projection on a plane by applying an "arc to chord" correction, i.e., the $(t - T)$ correction. Figure 11–2 illustrates this conversion for a Transverse Mercator projection.

In the figure,

$$T = \text{observed azimuth referred to grid north,}$$
$$t = \text{plane azimuth,}$$
$$t - T = k_t\,(Y_A - Y_B)\,\frac{2X_A + X_B}{3},$$

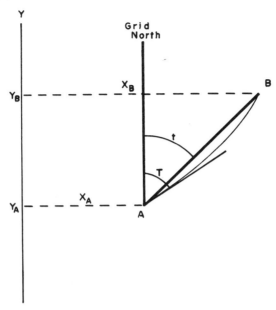

Figure 11–2. Arc to chord.

where $k_t = f(\phi)$ is a function of geodetic latitude; for example:

ϕ	$k_t \times 10^{-10}$ sec/meters^2
0°	25.52
30°	25.44
45°	25.35
60°	25.26
90°	25.18

Observed directions referred to grid north, T, may thus be reduced to a direction on the plane, t, consistent with the plane rectangular coordinates.

For the same reasons, observed distances should also be reduced to the plane. The theory of both direction and distance reductions is contained in [1].

On this computational plane, grid azimuth may be written

$$\tan t = \frac{X_B - X_A}{Y_B - Y_A},$$

or

$$(Y_B - Y_A) \tan t = X_B - X_A;$$

and, by differentiation,

$$\sec^2 t \, dt (\Upsilon_B - \Upsilon_A) + \tan t (d\Upsilon_B - d\Upsilon_A) = dX_B - dX_A,$$

$$dt = \cos^2 t \, \frac{dX_B - dX_A}{\Upsilon_B - \Upsilon_A} - \cos t \sin t \, \frac{d\Upsilon_B - d\Upsilon_A}{X_B - X_A}.$$

Noting that

$$\sin t = \frac{X_B - X_A}{S} \qquad \text{and} \qquad \cos t = \frac{\Upsilon_B - \Upsilon_A}{S},$$

where S is the distance AB, the differential equation now may be written

$$dt = \frac{\Upsilon_B - \Upsilon_A}{S^2} (dX_B - dX_A) - \frac{X_B - X_A}{S^2} (d\Upsilon_B - d\Upsilon_A). \quad (11\text{--}29)$$

This equation represents changes in a computed azimuth in radians as a function of coordinate changes. It is the classical first step toward an observation equation.

Amending the symbols used previously, we now have

$$v_{ij} = L_{ij} - (l_{ij} + z_i),$$

and $z_i =$ the station correction for station i, the observing station. The subscript j will represent the station observed, and v designates the correction to observations used to obtain consistency. These corrections are independent as are the individual observations. However, for directions to have meaning, they must be referred to an initial quantity—an initial sighting. An error in this initial assumption affects all subsequent pointings by an equal amount which is in addition to the random error of sighting. There is a z correction for every station from which directions are observed. The situation is depicted in Fig. 11–3 where l_1 and l_2 represent observed grid azimuths and l'_1 and l'_2 are these azimuths corrected for error in initial orientation—

$$l' = l + z.$$

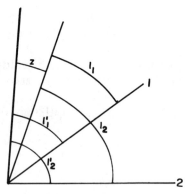

Figure 11–3. z correction.

The random error portion of v must still be accommodated.

The correction, dt, to the computed observation must result in the same quantity as the correction to the actual observation, that is,

$$l_{ij}^0 + dt = L_{ij} = l_{ij} + v_{ij} + z_i;$$

therefore, Eq. (11–29) may now be molded into a general observation equation for directions:

$$v_{ij} = -z_i + a_{ij} (x_j - x_i) + b_{ij} (y_j - y_i) - l_{ij}.$$

The x and y are corrections to station coordinates, replacing the dX and dY terms of Eq. (11–29), and with z are the unknowns to be determined. The constant coefficients are

$$a_{ij} = \frac{Y_j - Y_i}{S_{ij}^2} \rho'' \quad \text{and} \quad b_{ij} = -\frac{X_j - X_i}{S_{ij}^2} \rho''$$

where ρ'' is the arc-second equivalant to one radian.

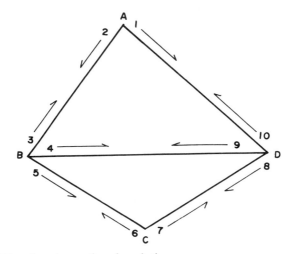

Figure 11–4. Direction observations for solution.

Consider applying this equation to the problem shown in Fig. 11–4. The A, B, and C are known; the position of D is desired. All stations are occupied and all lines are observed. There are ten knowns and six unknowns—four station corrections, and two coordinate corrections. When both stations of a line are known, x and y are zero and the equation only contains v, z, and l. For reciprocal observations along the same line, the coefficients a and b have the same sign. The fact that

$$a_{ADX_D} = -a_{DA} (-x_D) = a_{DA}x_D$$

may be seen by referring to Eq. (11–29) and reversing the direction of the line:

$$v_{AD} = -z_A + a_{AD}x_D + b_{AD}y_D - l_{AD}$$
$$v_{AB} = -z_A \qquad\qquad\qquad\qquad - l_{AB}$$
$$v_{BA} = -z_B \qquad\qquad\qquad\qquad - l_{BA}$$
$$v_{BD} = -z_B + a_{BD}x_D + b_{BD}x_D - l_{BD}$$
$$v_{BC} = -z_B \qquad\qquad\qquad\qquad - l_{BC}$$
$$v_{CB} = -z_C \qquad\qquad\qquad\qquad - l_{CB}$$
$$v_{CD} = -z_C + a_{CD}x_D + b_{CD}y_D - l_{CD}$$
$$v_{DC} = -z_D + a_{CD}x_D + b_{CD}y_D - l_{DC}$$
$$v_{DB} = -z_D + a_{BD}x_D + b_{BD}y_D - l_{DB}$$
$$v_{DA} = -z_D + a_{AD}x_D + b_{AD}y_D - l_{DA}.$$

The individual observation equations are squared and summed to yield $[v^2]$. The expression is differentiated with respect to the six corrections to form six normal equations. These, in turn, are solved for the unknowns.

If the z terms could be removed, the number of unknowns requiring solution would be greatly reduced. There are several methods of doing this, such as the Schreiber method [3]. By far the most direct approach to this problem is to consider observation of angles rather than directions. Observation equations for angles may be formed by subtracting adjacent direction equations, thereby cancelling the z terms:

$$v_{AD} - v_{AB} = v_{DAB} = a_{AD}x_D + b_{AD}y_D - l_{DAB}$$
$$v_{BA} - v_{BD} = v_{ABD} = a_{BD}x_D + b_{BD}y_D - l_{ABD}$$
$$v_{BD} - v_{BC} = v_{DBC} = a_{BD}x_D + b_{BD}y_D - l_{DBC}$$
$$v_{BC} - v_{CD} = v_{BCD} = a_{CD}x_D + b_{CD}y_D - l_{BCD}$$
$$v_{DC} - v_{DB} = v_{CDB} = a_{CB}x_D + b_{CB}y_D - l_{CDB}$$
$$v_{DB} - v_{DA} = v_{BDA} = a_{BA}x_D + b_{BA}y_D - l_{BDA}.$$

The $l_{DAB} = l_{AD} - l_{AB}$, etc. Again referring to Eq. (11–29), the coefficients of the last two equations are seen to be

$$(a, b)_{CB} = (a, b)_{CD} - (a, b)_{BD},$$

and

$$(a, b)_{BA} = (a, b)_{BD} - (a, b)_{AD}.$$

There are now six observation equations which may be normalized and solved for the two unknowns.

The corrections to $X_D{}^0$ and $Y_D{}^0$ are made and Eq. (11–22) will yield the $[v^2]$. Equation (11–23) defines the standard error of unit weight,

$$m = \left(\frac{[v^2]}{4}\right)^{1/2}.$$

With m and Eqs. (11–25) and (11–26), the error estimates for the adjusted coordinates are computed.

In the preceding solutions, distance observations and direction or angle observations have been treated separately. However, in practice, such observations are combined. One way to effect such a combination is to express all equations in the same units. The values of v must be in either angular or linear units and relative weights must be properly assigned to the different measurement types.

Suppose, for instance, that the length of line BD was observed in the scheme of Fig. 11–4. This observation should be included in the solution array. Since the existing equations are in angular units, this distance observation should likewise be so expressed. Observation Eq. (11–17) for this distance measurement should be expressed as

$$v_{BD} = a_{BD} x_D + b_{BD} y_D - l_{BD}$$

where

$$a_{BD} = \frac{X_D - X_B}{S_{BD}} \quad \text{and} \quad b_{BD} = \frac{Y_D - Y_B}{S_{BD}}.$$

Of course, the v, l, x, and y are in linear units. If this equation is multiplied by ρ''/S, it is transformed to angular units. Thus converted, the l and v may be considered short arcs of radius S_{BD}, that is, $\rho'' v/S$ is the angle subtending arc v. With this multiplication,

$$v'' = a'' x + b'' y - l''$$

where

$$a'' = \frac{X_D - X_B}{S^2{}_{BD}} \rho'' \quad \text{and} \quad b'' = \frac{Y_D - Y_B}{S^2{}_{BD}} \rho''.$$

The corrections to be determined, x and y, are still in linear units, however.

The next step is to determine the weight and the contribution of this observation to the solution. First, for the angles, Eq. (11–9) is applied to the angle observation being used. If the standard error of a direction is known from past experience and is signified with m, then that of the angle resulting from such a direction is

$$m_{\angle} = (2)^{1/2} m.$$

Now, for the distance measurements, the standard error in linear units, m_d, is converted to arc seconds with multiplication by ρ''/S,

$$m''{}_d = \frac{\rho''}{S} m_d.$$

Recall Eq. (11–14) which relates m and weight, p;

$$\frac{p_1}{p_2} = \frac{m_2{}^2}{m_1{}^2}.$$

With this expression, the relative weight between the angle and distance equations is determined as

$$\frac{p_d}{p_\chi} = \frac{m^2_\chi}{m''^2_d} = \frac{S^2 m^2_\chi}{(\rho'' m_d)^2}.$$

Assuming the $m_\chi > m''_d$, it is convenient to take $p_\chi = 1$, and the weight of the distance equation becomes

$$p_d = \frac{S^2 m^2_\chi}{(\rho'' m_d)^2}.$$

Because this weight is a function of S, it is apparent that the value assigned will vary with lines of different length if m_d is constant. Normally, however, the ratio S/m_d will be reasonably constant over the measured distances used.

The distance observation equation is multiplied by $(p_d)^{1/2}$, the dimensionless weight, i.e.,

$$(p_d)^{1/2} v'' = (p_d)^{1/2} a''x + (p_d)^{1/2} b''y - (p_d)^{1/2} l'',$$

and is then included in the adjustment array.

In the same way, angle observation equations may be modified for compatability with an array of distance equations.

The steps employed with "variation of parameters" will be summarized with a numerical example after a look at "conditions," the next subject.

Conditions. In adjustment by this technique, the observations are altered to conform to a mathematical condition or model. In the solution, there is one unknown per condition. While "variation of parameters" produced a coordinate adjustment directly, this method adjusts the observations and these values then are used to compute a consistent position for the unknown station.

The difference between the number of observations, n, or knowns, and the unknowns, U, defines the number of conditions, r, that is,

$$r = n - U = \text{conditions}.$$

If $r < U$, solution by conditions is said to be simpler than by variation of parameters because the number of unknowns is reduced. However, this has become hardly more than an academic viewpoint because of the use of electronics in computations.

Symbolize condition a to be satisfied in the survey adjustment by

$$a_0 + a_1 L_1 + a_2 L_2 + \ldots a_n L_n = 0. \tag{11-30}$$

This means that the adjusted observations, L_i, must satisfy this model. Of course, with the measurements, l_i,

$$a_0 + a_1 l_1 + a_2 l_2 + \ldots a_n l_n = W_a; \tag{11-31}$$

condition a is not satisfied by a misclosure, W_a, due to observation errors. With $L_i = l_i + v_i$ substituted in Eq. (11–30),

$$a_0 + a_1(l_1 + v_1) + a_2(l_2 + v_2) + \ldots a_n(l_n + v_n) = 0$$

is produced. This expression upon expansion and further substitution from Eq. (11–31) is written for conditions a through r:

$$
\begin{aligned}
a_1 v_1 + a_2 v_2 + \ldots a_n v_n + W_a &= 0 \\
b_1 v_1 + b_2 v_2 + \ldots b_n v_n + W_b &= 0 \\
c_1 v_1 + c_2 v_2 + \ldots c_n v_n + W_c &= 0 \\
\cdots \quad \cdots \quad \cdots \quad \cdots \quad \cdots \\
r_1 v_1 + r_2 v_2 + \ldots r_n v_n + W_r &= 0.
\end{aligned}
\tag{11–32}
$$

There are r such *condition equations* written in terms of corrections, v_i, to n observations. Similar to the observation equation in variation of parameters, the coefficients are partial derivatives, i.e.,

$$a_i, b_i, c_i \ldots = \frac{\partial f_a}{\partial l_i}, \frac{\partial f_b}{\partial l_i}, \frac{\partial f_c}{\partial l_i} \ldots$$

In this case, however, these are partials of the conditions in terms of observations.

At this point, some of the typical conditions encountered in geodesy are described to clarify the discussion.

Angle conditions (plane angles)

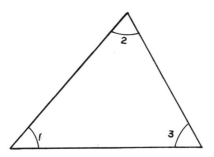

Figure 11–5. Angle conditions for a triangle.

For a triangle, Fig. 11–5, the condition is

$$-180° + L_1 + L_2 + L_3 = 0.$$
$$\text{Misclosure:} \quad -180° + l_1 + l_2 + l_3 = W.$$
$$\text{Condition equation:} \quad v_1 + v_2 + v_3 + W = 0,$$

where the v_i terms are corrections to the observed angles. Obviously, the partial derivatives are unity, that is, $a_1, a_2, a_3 = 1$.

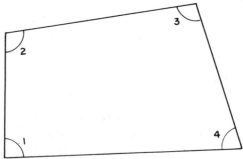

Figure 11–6. Angle conditions for a rectangle.

For a rectangle, Fig. 11–6, the condition is

$$-360° + L_1 + L_2 + L_3 + L_4 = 0.$$

The misclosure and condition equation are obvious.

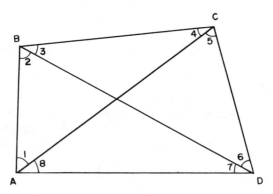

Figure 11–7. Conditions for a quadrilateral.

For a quadrilateral, Fig. 11–7, the conditions are:

$$-180° + L_1 + L_2 + L_3 + L_4 = 0,$$
$$-180° + L_3 + L_4 + L_5 + L_6 = 0,$$
$$-180° + L_5 + L_6 + L_7 + L_8 = 0.$$

Three triangle misclosures:

$$W_a, \quad W_b, \quad W_c.$$

Three condition equations:

$$v_1 + v_2 + v_3 + v_4 + W_a = 0,$$
$$v_3 + v_4 + v_5 + v_6 + W_b = 0,$$
$$v_5 + v_6 + v_7 + v_8 + W_c = 0.$$

(Note that all observed angles have been used; therefore, triangle
ABD cannot be used for a fourth condition.)

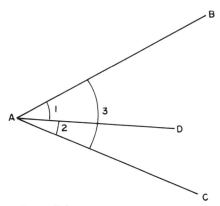

Figure 11–8. Fixed angle conditions.

For a fixed angle, Fig. 11–8 (A, B, C are fixed stations), the condition is:

$$-L_3 + L_1 + L_2 = 0.$$
$$\text{Misclosure: } -L_3 + l_1 + l_2 = W.$$
$$\text{Condition equation: } v_1 + v_2 + W = 0.$$

(The same relationship would hold for a fixed azimuth between known stations in direction observations.)

Side condition, Fig. 11–7.

To establish this condition, note from the law of sines that

$$AB = \frac{\sin l_4}{\sin l_1} BC,$$

$$BC = \frac{\sin l_6}{\sin l_3} CD,$$

$$CD = \frac{\sin l_8}{\sin l_5} DA,$$

$$DA = \frac{\sin l_2}{\sin l_7} AB.$$

From this it is apparent that

$$AB = \frac{\sin l_4 \sin l_6 \sin l_8 \sin l_2}{\sin l_1 \sin l_3 \sin l_5 \sin l_7} AB = Q(AB);$$

therefore, the condition is

$$Q = 1.$$
$$\text{Misclosure: } (Q-1)\rho'' = W.$$

Multiplying by ρ'' converts the misclosure to seconds of arc for inclusion with angular conditions. To derive the coefficients and the side-condition equation, consider triangle ABC. Now

$$AB = \frac{\sin l_4}{\sin l_1} BC;$$

and differentiating,

$$d(AB) = \frac{BC \, (\sin l_1 \cos l_4 \, dl_4 - \sin l_4 \cos l_1 dl_1)}{\sin^2 l_1} + \frac{\sin l_4}{\sin l_1} d(BC).$$

Noting that $\sin l_4/\sin l_1 = AB/BC$, this differential expression may be written

$$d(AB) = BC \left(\frac{\sin l_4 \dfrac{\cos l_4}{\sin l_4}}{\sin l_1} dl_4 - \frac{\sin l_4}{\sin l_1} \frac{\cos l_1}{\sin l_1} dl_1 \right) + \frac{AB}{BC} d(BC),$$

$$= BC \frac{\sin l_4}{\sin l_1} (\cot l_4 dl_4 - \cot l_1 dl_1) + \frac{AB}{BC} d(BC),$$

$$\frac{d(AB)}{AB} = \frac{d(BC)}{BC} + \cot l_4 dl_4 - \cot l_1 dl_1.$$

A similar expression may be written for

$$\frac{d(BC)}{BC} = f \left(\frac{d(CD)}{CD} , l_6, l_3 \right)$$

and

$$\frac{d(CD)}{CD} = f \left(\frac{d(DA)}{DA} , l_8, l_5 \right)$$

etc. When these are substituted, in turn, in the $d(AB)/AB$ expression and with $v_i = dl_i$, the side-condition equation is formed.

Condition equation:

$$v_4 \cot l_4 + v_6 \cot l_6 + v_8 \cot l_8 + v_2 \cot l_2 - v_1 \cot l_1 - v_3 \cot l_3$$
$$-v_5 \cot l_5 - v_7 \cot l_7 + W = 0.$$

Interestingly, a side has not been measured for this side condition. If the survey connects fixed sides, the same type of equation is formed and the misclosure, W, is now the difference between the fixed value and the value computed with observed values.

Many other interesting conditions are possible and the challenged reader is referred to [1] – [3].

Now, what is to be done with these conditions? As in the previous method discussed, corrections to the observations are desired such that

$$[v^2] = v_1{}^2 + v_2{}^2 + v_3{}^2 + \ldots v_n{}^2 = \text{minimum}.$$

Although corrections to observations, v_i, are sought, these are not obtained at once. A solution for an unknown related to each condition is first pursued.

Each condition equation (equations a through r) is multiplied by a term called a correlate, $-k$. This k is the unknown to be determined. Referring to Eqs. (11–32), multiply equation a by $-k_a$, b by $-k_b$, etc. Once done, add these equations to $[v^2]/2$. After collecting terms common to v_i, this summation is written:

$$K = [v^2]/2 - v_1(k_a a_1 + k_b b_1 + \ldots k_r r_1) - v_2(k_a a_2 + k_b b_2 + \ldots k_r r_2)$$
$$- \ldots v_n(k_a a_n + k_b b_n + \ldots k_r r_n).$$

In the above equation when K is a minimum, so must $[v^2]$ be. Therefore, differentiate K relative to the v_i terms:

$$dK = \frac{\partial K}{\partial v_i} \, dv_i = (-k_a a_1 - k_b b_1 - \ldots k_r r_1 + v_1) \, dv_1$$
$$+ (-k_a a_2 - k_b b_2 - \ldots k_r r_2 + v_2) \, dv_2$$
$$+ \ldots (-k_a a_n - k_b b_n - \ldots k_r r_n + v_n) \, dv_n.$$

Now dK would equal zero and K and $[v^2]$ would have minimum values providing the coefficient expressions for the dv_i terms were zero, that is,

$$-k_a a_i - k_b b_i - \ldots k_r r_i + v_i = 0.$$

Therefore, determine k_i such that

$$v_1 = a_1 k_a + b_1 k_b + \ldots r_1 k_r$$
$$v_2 = a_2 k_a + b_2 k_b + \ldots r_2 k_r \qquad (11\text{–}33)$$
$$\ldots \quad \ldots \quad \ldots \quad \ldots \quad \ldots$$
$$v_n = a_n k_a + b_n k_b + \ldots r_n k_r.$$

If the observations represented by the v_i have different relative weights, each term on the right must be divided by p_i.

Substitution of these expressions for v_i in the condition equations (11–32) yields normal equations to be solved for the k terms:

$$[a^2]k_a + [ab]k_b + \ldots [ar]k_r + W_a = 0$$
$$[ab]k_a + [b^2]k_b + \ldots [br]k_r + W_b = 0 \qquad (11\text{–}34)$$
$$\ldots \quad \ldots \quad \ldots \quad \ldots \quad \ldots \quad \ldots$$
$$[ar]k_a + [br]k_b + \ldots [r^2]k_r + W_r = 0.$$

With the values of k thus determined and substituted in Eqs. (11–33), the corrections, v_i, are computed; these, in turn, with

$$L_i = l_i + v_i$$

yield adjusted values for the observations which are compatible with the conditions imposed. The adjusted observations are then used to compute the coordinates of the unknown point. Since all the observations are now consistent, only the minimum need be considered in the position computation. For instance, suppose three directions from three known points intersected at the point to be determined. Only two of these need be employed in the computation.

As the individual corrections, v_i, were determined, they may be squared and summed to yield $[v^2]$. Or, similar to Eq. (11–22), by squaring Eqs. (11–33), summing, and referring to Eqs. (11–34), it will be found that

$$[v^2] = -k_a W_a - k_b W_b - \ldots k_r W_r = -[kW]. \qquad (11\text{–}35)$$

Now, by Eq. (11–23), the standard error of unit weight is

$$m = \left(\frac{[v^2]}{n-U}\right)^{1/2} = \left(\frac{[v^2]}{r}\right)^{1/2}.$$

For observations where the weight is not unity, equations of the forms of Eq. (11–25) or Eq. (11–26) may be employed.

Example. This discussion of least squares will be concluded with an

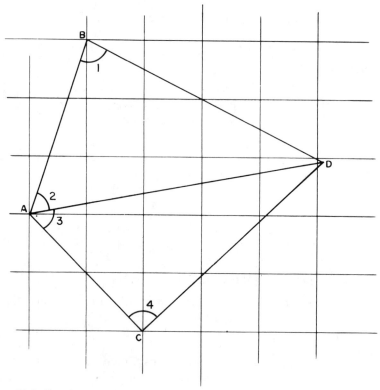

Figure 11–9. Graph paper survey.

example adjustment. The steps for "variation of parameters" and "conditions" are summarized with the graph paper survey shown in Fig. 11–9. In order not to cloud the illustration with unnecessary mathematical rigor, all computations are on a plane XY coordinate system, five-place trigonometric functions are employed, and angles are recorded in minutes of arc (in contrast to the eight-place tables and seconds of arc used in practice).

The known stations and fixed coordinates are:

Station	X	Y
A	1000	3000
B	2000	6000
C	3000	1000

From the fixed coordinates, the following azimuths are computed:

$$\alpha_{AB} = 18°26'.09,$$
$$\alpha_{BA} = 198°26'.09,$$
$$\alpha_{AC} = 135°00'.00,$$
$$\alpha_{CA} = 315°00'.00;$$

also, fixed angle,

$$\sphericalangle CAB = \alpha_{AC} - \alpha_{AB} = 116°33'.91;$$

and fixed sides,

$$S_{AB} = 3162.278$$
$$S_{AC} = 2828.427.$$

The angle *observations* are:

$$l_1 = \sphericalangle ABD = 82°20'$$
$$l_2 = \sphericalangle DAB = 60°36'$$
$$l_3 = \sphericalangle CAD = 55°58'$$
$$l_4 = \sphericalangle DCA = 91°04'.$$

This is the *given* information for both adjustment techniques.

Variation of Parameters

1. Select approximate coordinates for D:

$$X^0{}_D = 6105; \quad Y^0{}_D = 3990.$$

2. With these approximate coordinates, compute azimuths from the known to the unknown stations—$\tan \alpha = \Delta X / \Delta Y$;

$$\alpha^0{}_{BD} = 116°05'.33,$$
$$\alpha^0{}_{AD} = 79°01'.50,$$
$$\alpha^0{}_{CD} = 46°04'.93.$$

Then compute the angles that correspond to those observed:

$$l_1{}^0 = \alpha_{BA} - \alpha^0{}_{BD} = \sphericalangle^0{}_{ABD} = 82°20'.76,$$
$$l_2{}^0 = \alpha^0{}_{AD} - \alpha_{AB} = \sphericalangle^0{}_{DAB} = 60°35'.41,$$
$$l_3{}^0 = \alpha_{AC} - \alpha^0{}_{AD} = \sphericalangle^0{}_{CAD} = 55°58'.50,$$
$$l_4{}^0 = \alpha^0{}_{CD} - \alpha_{CA} = \sphericalangle^0{}_{DCA} = 91°04'.93.$$

Also,

$$S^{02}{}_{BD} = 20.8911 \times 10^6,$$
$$S^{02}{}_{AD} = 27.0411 \times 10^6$$
$$S^{02}{}_{CD} = 18.5811 \times 10^6.$$

3. Determine the observed minus computed values, $l_i = l_i - l_i{}^0$:

$$l_1 = -0'.76,$$
$$l_2 = 0'.59,$$
$$l_3 = -0'.50,$$
$$l_4 = -0'.93.$$

4. Compute coefficients a and b for observation equations. For directions,

$$v_{ij} = -z_i + a_{ij}(x_j - x_i) + b_{ij}(y_j - y_i) - l_{ij},$$

where the coefficients are

$$a_{ij} = \frac{Y_j - Y_i}{S^2{}_{ij}} \rho'' \qquad \text{and} \qquad b_{ij} = \frac{X_j - X_i}{S^2{}_{ij}} \rho''.$$

For angle observation equations, those for adjacent directions are subtracted. Furthermore, recall that

$$l_1 = \sphericalangle_{ABD} = \alpha_{BA} - \alpha_{BD}.$$

As the line BA is fixed, the coordinate corrections, x and y, appear only in the direction equation for BD. In this case, the coefficients are negative. The sign of the coefficients depends on the manner in which l_i is computed. In this example, angles are computed right direction minus left direction. Also, in keeping with observations recorded in minutes of arc, ρ' replaces ρ''; $\rho' = 3437.75$.

$$a_1 = -a_{BD} = 0.33076; \qquad\qquad b_1 = -b_{BD} = 0.67550$$
$$a_2 = a_{AD} = 0.12586; \qquad\qquad b_2 = b_{AD} = -0.64900$$
$$a_3 = -a_2 = -a_{AD} = -0.12586; \qquad b_3 = -b_2 = -b_{AD} = 0.64900$$
$$a_4 = a_{CD} = 0.55319; \qquad\qquad b_4 = b_{CD} = -0.57447$$

5. Write the observation equations, $v_i = a_i x + b_i y - l_i$:

$$v_1 = 0.33076\,x + 0.67550\,y + 0.76,$$
$$v_2 = 0.12586\,x - 0.64900\,y - 0.59,$$
$$v_3 = -0.12586\,x + 0.64900\,y + 0.50,$$
$$v_4 = 0.55319\,x - 0.57447\,y + 0.93.$$

6. Form and solve the normal equations,

$$[a^2] x + [ab]y - [al] = 0,$$
$$[ab] x + [b^2]y - [bl] = 0:$$
$$0.44710 x - 0.25773 y + 0.62866 = 0,$$
$$-0.25773 x + 1.62872 y + 0.68654 = 0;$$

from which

$$x = -1.8146 \quad \text{and} \quad y = -0.7086.$$

7. Compute the adjusted coordinates:

$$X_D = X^0{}_D + x = 6103.18 \quad \text{and} \quad Y_D = Y^0{}_D + y = 3989.29.$$

8. Compute the standard error of unit weight. First, the sum of the residuals squared is required. Substituting the x and y values from step 6 in the observation equations of step 5 yields

$$v_1 = -0'.32,$$
$$v_2 = -0'.36,$$
$$v_3 = 0'.27,$$
$$v_4 = 0'.33;$$

from which $[v^2] = 0'.41$. A like quantity is computed with Eq. (11–22). With Eq. (11–23),

$$m = \left(\frac{[v^2]}{n - U}\right)^{1/2} = \left(\frac{0.41}{2}\right)^{1/2} = \pm 0'.45.$$

9. Compute the standard error of the adjusted coordinates. From

$$p_x = \frac{[a^2]\,[b^2] - [ab]^2}{[b^2]} = 0.41$$

and Eq. (11–25),

$$m_x = \frac{m}{(p_x)^{1/2}} = \pm 0.70,$$

which is the standard error of X_D. Likewise, for Y,

$$p_y = \frac{[a^2]\,[b^2] - [ab]^2}{[a^2]} = 1.48$$

and from Eq. (11–26),

$$m_y = \frac{m}{(p_y)^{1/2}} = \pm 0.37.$$

Conditions

1. Determine the conditions to be satisfied and the misclosures existing with the observations. There are

$$r = n - U = 4 - 2 = 2 \text{ conditions.}$$

a. For the fixed angle condition, designated condition a for this adjustment,

$$-L_{CAB} + L_{DAB} + L_{CAD} = 0.$$

Therefore the misclosure is

$$W_a = -L_{CAB} + l_2 + l_3 = 0'.09.$$

b. For the side condition, condition b, consider

$$AD = AB \frac{\sin L_{ABD}}{\sin L_{BDA}}$$

in triangle ABD; and

$$AD = AC \frac{\sin L_{DCA}}{\sin L_{ADC}}$$

in triangle ACD. As $AD = AD$, the condition is

$$\frac{AB}{AC} = \frac{\sin L_{DCA} \sin L_{BDA}}{\sin L_{ADC} \sin L_{ABD}}.$$

The misclosure is

$$W_b = \left(\frac{\sin l_4 \sin l_5}{\sin l_6 \sin l_1} - \frac{AB}{AC} \right) \rho' = -1'.92,$$

where

$$l_5 = 180° - l_1 - l_2 = 37° \ 04'$$

and

$$l_6 = 180° - l_3 - l_4 = 32° \ 58'.$$

2. Write the condition equations, Eqs. (11–32);

$$a_1 v_1 + a_2 v_2 + \ldots a_n v_n + W_a = 0$$
$$b_1 v_1 + b_2 v_2 + \ldots b_n v_n + W_b = 0, \text{ etc.}$$

a. For the fixed angle condition,

$$v_2 + v_3 + 0'.09 = 0.$$

b. For the side condition, recalling the general form

$$v_4 \cot l_4 + v_5 \cot l_5 - v_1 \cot l_1 - v_6 \cot l_6 + W_b = 0,$$

and noting that

$$v_5 = -v_1 - v_2 \quad \text{and} \quad v_6 = -v_3 - v_4;$$

thus

$$1.52318 \, v_4 + 1.54180 \, v_3 - 1.45841 \, v_1 - 1.32380 \, v_2 - 1'.92 = 0.$$

3. Form correlate equations (11–33) for each observation,

$$v_1 = a_1 k_a + b_1 k_b$$
$$v_2 = a_2 k_a + b_2 k_b, \text{ etc.:}$$

$$v_1 = \qquad -1.45841 \, k_b$$

$$v_2 = k_a \quad - 1.32380\, k_b$$
$$v_3 = k_a \quad + 1.54180\, k_b$$
$$v_4 = \qquad + 1.52318\, k_b.$$

4. Form and solve the normal equations (11–34),

$$[a^2]k_a + [ab]k_b + W_a = 0$$
$$[ab]k_a + [b^2]k_b + W_b = 0:$$
$$2.00000\, k_a + 0.21800\, k_b + 0.09 = 0$$
$$0.21800\, k_a + 8.57663\, k_b - 1.92 = 0;$$
$$k_a = -0.06959 \qquad \text{and} \qquad k_b = 0.22563.$$

5. Substitute k_a and k_b in the correlate equations of step 3 and determine the residuals or corrections to the observations:

$$v_1 = -0'.33$$
$$v_2 = -0'.37$$
$$v_3 = \quad 0'.28$$
$$v_4 = \quad 0'.34.$$

6. Correct the observed angles, $L_i = l_i + v_i$:

$$L_{ABD} = 82°19'.67$$
$$L_{DAB} = 60°35'.63$$
$$L_{CAD} = 55°58'.28$$
$$L_{DCA} = 91°04'.34.$$

7. With the adjusted observations, compute the coordinates for D. A simple way is to intersect from any two of the three fixed stations (all three intersection problems should yield identical results with the adjusted observations). Using B and C, compute graph azimuth α_{BD} and α_{CD}:

$$\alpha_{BD} = \alpha_{BA} - L_{ABD} = 116°06'.42 \text{ and } \alpha_{CD} = \alpha_{CA} + L_{DCA} = 46°04'.34.$$

With $\tan\alpha = \varDelta X / \varDelta Y$, the following expressions may be written and solved:

$$Y_D = \frac{(X_C - X_B) + Y_B \tan\alpha_{BD} - Y_C \tan\alpha_{CD}}{\tan\alpha_{BD} - \tan\alpha_{CD}} = 3989.31$$

$$X_D = \frac{(Y_C - Y_B)\tan\alpha_{BD}\tan\alpha_{CD} + X_B \tan\alpha_{CD} - X_C \tan\alpha_{BD}}{\tan\alpha_{CD} - \tan\alpha_{BD}} = 6103.21.$$

8. Compute the standard error of unit weight. By squaring and summing the values of v_i from step 5,

$$[v^2] = 0'.44.$$

Also, from Eq. (11–35),

$$[v^2] = -[kW] = 0'.44.$$

With Eq. (11–23),

$$m = \left(\frac{[v^2]}{r}\right)^{1/2} = \left(\frac{0'.44}{2}\right)^{1/2} = \pm\, 0'.47.$$

Comparison of the adjusted coordinates determined in step 7 and the residuals produced by both methods shows close agreement. Theoretically, the results should be identical. Disregarding discrepancies induced by truncation, the product of variation of parameters will agree with that of conditions if the coordinate adjustment does not exceed 1/5000 relative to the distance between known and unknown stations. The selection of approximate coordinates for the unknown station cannot be haphazard. If the correction exceeds this limit, the adjustment should be done again with value determined from the first try as the approximate starting coordinates.

ERROR PROBABILITY

The problem of *too* many answers has been solved. Additionally, a statement of adequacy for the answer has been found. Now ask the question, "What is this standard error?"

The standard error of a value is an estimate within which the true error has a certain chance of falling—that is, with a given probability, the true error will not exceed the standard error. To grasp this proposition, the concept of probability must be introduced.

A pair of dice are the classic example used for introduction to probability. Essentially, probability states that out of so many opportunities, a given "happening" will occur. For instance, in the roll of one die, what are the chances that a 3 will turn up? The 3 is one of six possible numbers; therefore, the chance it will occur in a single throw is

$$1/6 = 0.166.$$

The chance of not rolling a 3 is expressed by

$$1 - 1/6 = 0.834.$$

The chance of rolling any number between 1 and 6 in a roll is

$$6/6 = 1;$$

and of rolling a number greater than 6,

$$0/6 = 0.$$

The probability of an occurrence falls between 0 and 1 or 0 per cent and 100 per cent.

The probability of a combined event is simply the sum of the probabilities of the individual participants. For instance, what is the probability that a 3 will appear in one throw of both dice? Simply

$$1/6 + 1/6 = 1/3.$$

The probability of two events happening simultaneously is the product of the individual probabilities. What is the chance of a 3 and 4 occurring in one roll?

$$(1/6) \ (1/6) = 1/36 = 0.0278.$$

There is a 2.78 per cent chance that a 7 will be made with a 3 and a 4 in one toss of the dice.

Because there are six possible combinations that make 7, $(3,4; 4,3; 2,5; 5,2; 1,6; 6,1)$, the total probability of making 7 in one roll is the sum of these individual probabilities, that is,

$$1/36 + 1/36 + 1/36 + 1/36 + 1/36 + 1/36 = 1/6.$$

The probability of other numbers being formed in one throw are less because there are fewer combinations available. For example, there are only four ways to make 5, three for 10, etc. This situation is illustrated graphically. With the numbers from 1 to 13 plotted along the horizontal, and the increments along the vertical being 1/36, the result is shown in

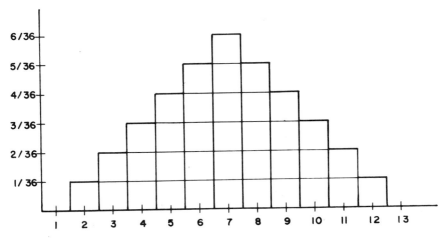

Figure 11–10. Probability.

Fig. 11–10. With the individual squares assigned an area of 1/36, the total area enclosed is seen to equal 1.

If the dice thrown in a single throw are increased in number, this bar graph approaches a smooth curve. Such a curve is a "*normal* probability density curve". This curve represents the total probability of an occurrence of a random event and encloses an area which is equal to 1, or 100 per cent. In the case of the two dice, this means that there is a 100 per cent probability that a number between 2 and 12 will result. This is the bell curve.

The equation for such a curve is written

$$y = \frac{1}{m \, (2\pi)^{1/2}} \, e^{-v^2/2 m^2};$$

where $v = L - l_i$, $L =$ mean value of l_i, $l_i =$ random variable, $m^2 = [v^2]/(n - U) =$ standard error, and $e =$ base of natural logarithms.
The specific l_i corresponding to the maximum y is L, or the mean of an infinite number of random l_i. The curve is symmetrical about this value, that is,

$$[-v] = [+v].$$

Since the true value is never known and an infinite number of measures is unlikely, the mean, L, and residual, v, must be accepted as for truth and error, respectively. Within these necessary assumptions, this curve is used to estimate errors and their probability.

The fact that the area under the curve represents 100 per cent means that there is a 100 per cent probability of all errors occurring. To determine the area under the curve bounded by specific values and, in turn, the probability of their occurrence, this function is merely integrated between this interval. Take $+m$ and $-m$, for instance,

$$\text{Probability} = \int_{-m}^{+m} y \, dv = \int_{-m}^{+m} \frac{1}{m \, (2\pi)^{1/2}} \, e^{-v^2/2 m^2} \, dv = 0.6827.$$

The area under the portion of the curve between $+m$ and $-m$ is 0.6827, which means that there is a 68.27 per cent chance this error will exist.

Referring again to the example of the kilometer line measured between the two posts, the

$$m = \pm \, 0.042 \text{ meter.}$$

This means that there is a 68.27 per cent chance that the true error of an observation is equal to or less than ± 0.042 meter. From the same example, the $M = \pm 0.017$ meter which, with the same probability, states that the true error of the mean will be equal to or less than 0.017 meter.

From the same set of observations, errors corresponding to different probabilities may be computed by changing the integration limits. For convenience, these values may be referred to the standard error. For instance, the probable error (50 per cent) is equal to $\pm 0.6745 \, m$. Based on the standard error, the conversion factors to errors of different probabilities are contained in Table 11–1.

TABLE 11-1

Error Conversion Factors: Error (%) = $K\,m$

Per cent	K (linear)	K (circular)
39.35		1.0000
50.00	0.6745	1.1774
68.27	1.0000	
90.00	1.6449	2.1460
99.00	2.5758	3.0349

All the rules for combining errors given by Eqs. (11–8) – (11–11) apply to error estimates of probability other than the standard error. As an example, for a sum, $C = A + B$;

$$\text{Error}_C\,(90\%) = [\,(\text{Error}_A\,(90\%))^2 + (\text{Error}_B\,(90\%))^2\,]^{1/2}.$$

One-dimensional error estimates in X and Y are often combined in a two-dimensional ellipsoidal or circular error. If

$$m_X = m_Y,$$

then the circular standard error, CSE, is

$$m_c = m_X = m_Y.$$

Recalling that the probability of two events occurring simultaneously is the product of the individual probabilities, the expression for this type of error distribution is obtained by multiplying respective y curves for X and Y error distribution and then integrating between the limits of $\pm m_c$. It is found that the area under this curve is 0.3935. This means that there is a 39.35 per cent probability that the error falls within or on a circle of radius m_c.

When m_X does not equal m_Y but their ratio falls within 1.0 and 0.6, m_c may be adequately approximated with

$$m_c = \frac{m_X + m_Y}{2}.$$

As in the case of linear errors, the circular error may be defined for different probabilities. These conversion factors are also listed in Table 11–1.

From the foregoing discussion it is apparent that an error should never be stated unless defined as to its type and confidence level (per cent). To say, "I know the length of this line to plus or minus 0.01 meter," doesn't really state the case. This is an omission frequently encountered.

In concluding this discussion of adjustment computations, the reader is asked to remember:

1. The theory of least squares adjustment, whether applied with variation of parameters or conditions, as well as all probability error judgments stemming from normal distribution, applies only to random errors.

2. A random error is a small error which is positive as often as it is negative. With an infinite number of observations containing only random errors, the arithmetic mean would be the truth. With a finite number of observations to which we are normally limited, the mean approximates the truth, hopefully within the error estimated by normal distribution.

3. These adjustments do not improve observations but merely modify all data to conform to a model. An absolutely errorless observation may be altered to conform to a pattern set by its less perfect cousins; however, there is normally no way to identify perfection.

REFERENCES

1. G. Bomford, *Geodesy*, The Clarendon Press, Oxford, Second Edition, 1962.
2. W. Jordan and O. Eggert, *Handbuch der Vermessungskunde*, Vol. I, J. B. Metzlersche Verlagsbuchhandlung, Stuttgart, 1948.
3. H. F. Rainsford, *Survey Adjustments and Least Squares*, Frederick Ungar Publishing Co., New York, 1958.

Appendix A Formulas of Spherical Trigonometry

Many formulas of spherical trigonometry are used throughout the study of geodesy and, therefore, have been referred to in this book.

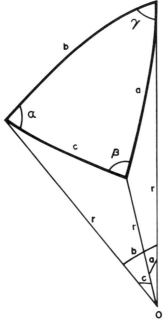

The accompanying sketch shows a spherical triangle with angles α, β, and γ. The sides a, b, and c, are expressed in angular units. The center of the sphere of radius r is designated O. The angles at this center corresponding to the sides are also a, b, and c and, of course, are identical in value to the sides.

The law of sines is:

$$\frac{\sin a}{\sin \alpha} = \frac{\sin b}{\sin \beta} = \frac{\sin c}{\sin \gamma}.$$

The law of cosines is:

$$\cos a = \cos b \cos c + \sin b \sin c \cos \alpha,$$
$$\cos b = \cos a \cos c + \sin a \sin c \cos \beta,$$
$$\cos c = \cos a \cos b + \sin a \sin b \cos \gamma.$$

The relationship between two angles and three sides is:

$$\sin c \cos \alpha = \cos a \sin b - \cos b \sin a \cos \gamma,$$
$$\sin a \cos \beta = \cos b \sin c - \cos c \sin b \cos \alpha,$$
$$\sin b \cos \gamma = \cos c \sin a - \cos a \sin c \cos \beta.$$

The relationship between three sides and an angle is:

$$\sin \frac{\alpha}{2} = \left(\frac{\sin (s-b) \sin (s-c)}{\sin b \sin c} \right)^{1/2},$$

$$\sin \frac{\beta}{2} = \left(\frac{\sin (s-c) \sin (s-a)}{\sin c \sin a} \right)^{1/2},$$

$$\sin \frac{\gamma}{2} = \left(\frac{\sin (s-a) \sin (s-b)}{\sin a \sin b} \right)^{1/2},$$

where $s = (a + b + c)/2$.

When one of the angles of the spherical triangle is equal to 90°, a right spherical triangle is formed to which the following equations pertain (assuming that γ is equal to 90° and the hypotenuse is c):

$$\sin b = \sin \beta \sin c,$$
$$\sin b = \tan a \cot \alpha,$$
$$\sin a = \sin \alpha \sin c,$$
$$\sin a = \tan b \cot \beta,$$
$$\cos c = \cos a \cos b,$$
$$\cos c = \cot \alpha \cot \beta,$$
$$\cos \beta = \cos b \sin \alpha,$$
$$\cos \beta = \tan a \cot c,$$
$$\cos \alpha = \cos a \sin \beta,$$
$$\cos \alpha = \tan b \cot c.$$

Appendix B Units of Distance Measurement

Effective July 1, 1959, new standards were adopted by the United States for the inch, the foot, the yard, and the nautical mile. These are known as international values and, with one exception, completely replace older values used in the United States. The exception is that, for geodetic work, the old United States foot (now called the American survey foot) will continue to be used until the basic survey networks of the United States are readjusted for the new international value of a foot.

The new standards align the United States with most other countries. However, until the use of the new values becomes general and the old values are completely forgotten, it is wise to identify the unit as international (or otherwise) wherever the difference is significant. Furthermore, it is quite possible that future years will bring changes; for this reason, it will be advisable for the geodesist to keep up to date on geodetic changes made by the U. S. Coast and Geodetic Survey and the U. S. Bureau of Standards (or comparable agencies in countries in which geodetic operations are conducted).

The currently accepted values are:

1 international inch = 25.4 millimeters exactly,
1 international foot = 0.3048 meter exactly,
1 American survey foot = 0.3048006 meter approximately,
1 Indian foot = 0.30479841 meter approximately,
1 meter = 3.2808398950 international feet approximately,
1 meter = 1.0936132983 international yards approximately,
1 international yard = 0.9144 meter exactly,
1 statute mile = 5280 international feet exactly,
1 statute mile = 1760 international yards exactly,
1 statute mile = 1609.344 meters exactly,
1 international nautical mile = 6076.10333 American survey feet approximately,
1 international nautical mile = 6076.115486 international feet approximately,
1 international nautical mile = 1852 meters exactly,
1 knot = 1 international nautical mile per hour.

Author Index

Numbers in parentheses indicate the numbers of the references when these are cited in the text without the names of the authors.

Numbers set in *italics* designate the page numbers on which the complete literature citation is given.

Subject Index